W9-AFM-113

The End of the Soviet Empire

By the same author

Réforme et Révolution chez les musulmans de l'Empire russe (Paris).
L'Empire éclaté (Paris: Flammarion, 1978)(Prix Aujourd'hui).
Lénine, la révolution et pouvoir (Paris: Flammarion, 1979).
Staline, l'ordre par la terreur (Paris: Flammarion, 1979).
Le Pouvoir confisqué (Paris: Flammarion, 1980).
Le Grand Frère (Paris: Flammarion, 1983).
La Déstalinisation commence (Paris: Complexe, 1986).
Ni paix ni guerre (Paris: Flammarion, 1986).
Le Grand Défi (Paris: Flammarion, 1987).
Le malheur russe. Essai sur le meurtre politique (Fayard, 1988).

The End of the Soviet Empire

The Triumph of the Nations

Hélène Carrère d'Encausse

TRANSLATED BY FRANKLIN PHILIP

A New Republic Book
BasicBooks
A Division of HarperCollins*Publishers*

Designed by Ellen Levine

94 95 96 97 CC/CW 9 8 7 6 5 4 3 2 1

Library of Congress Cataloging-in-Publication Data
Carrère d'Encausse, Hélène.
[Gloire des nations. English]
The end of the Soviet Empire: the triumph of the nations/ Hélène Carrère d'Encausse; translated by Franklin Philip.
p. cm.
"A New Republic Book."
Includes bibliographical references and index.
ISBN 0–465–09812–6 (cloth)
ISBN 0–465–09818–5 (paper)
1. Soviet Union—Politics and government—1985–1991. I. Title.
DK288.C3713 1992
947.085'4 — dc20
91–59006
CIP

Comrades, we are entitled to say that we have settled the nationalities question. The revolution paved the way for equality of rights among the national groups not only on the legal level but also on the socioeconomic level, and made a notable contribution to equalizing the economic, social, and cultural levels of development of all the republics and regions and of all the peoples. The friendship among the Soviet peoples is one of the greatest triumphs of the October Revolution. In itself, this is a unique phenomenon in world history and, for us, one of the fundamental pillars of the power and solidity of the Soviet state.

—Mikhail Gorbachev, November 2, 1987

It is vital to remember that once, before the sweep of Russian history was halted [by the revolution of 1917], we were a normal people and knew how to tell good from evil, how to see and hear.

—Alexander Tsipko, *Moskovskie Novosti*

Introduction

In December 1991, the Soviet Union officially expired.

- December 8: Russia, Ukraine, and Belarus created the Commonwealth of Slavic States in Minsk and declared the end of the union.
- December 17: Boris Yeltsin and Mikhail Gorbachev announced the USSR's official dissolution as of December 31, sixty-nine years after its founding.
- The Soviet Union was replaced by the Commonwealth of Independent States in which the republics of the former USSR rejoined the Slavic community.

Russia dominates this new group of former Soviet republics, but if, after dismantling communism, it is unable to ease the the long-suppressed conflicts and hatreds between the nations of the USSR, the victory will not bring forth democracy. That is the challenge Gorbachev's USSR was unable to meet and the one facing Russia today.

Can Russia respond successfully? One may as well ask whether it is a great modern nation or merely an old one again sliding into barbarism. To answer that momentous question, we must first try to understand how the vast Soviet Empire, the last empire on earth, crumbled before the world's astonished eyes.

When sometime during the 1970s, the dissident Andrei Amalrik asked whether the Soviet Union would survive until 1984,[1] his question was greeted in the West with polite or amused surprise. In his own country, the reaction was silence, the oblivion reserved for "traitors" who have lost their right to existence.

Nonetheless, the question seemed odd. The USSR was a superpower, the equal of the United States and recognized as such by the international community at Helsinki in 1975. It was also an imperial power, twice over. The empire of the czars, broken up by Lenin and World War I, then reconstructed by the same Lenin, covered 13 million square miles and included some 260 million inhabitants. An empire encompassing numerous ethnic groups within its borders, it prided itself on its status as a new kind of state, the state of all the people, in which a new historical community existed, the Soviet people. With World War II and Stalin, the original empire expanded into Western Europe. Was the West robbed? No, replied the leaders of the USSR, this was merely the second stage of the revolution, a new area in human progress, the gradual expansion of the new historical community foreshadowed by the Soviet people, the communist people.

These events were taken as evidence that within the Soviet Union the time of nations was over.

Those who later led the USSR—Brezhnev and Kosygin—launched Soviet power on an assault on the distant countries of Africa and Asia, whose conquest was to take the form of revolutions. For the first time in history, the Soviet empire was not concentrated in a continuous area, but like all traditional empires, extended beyond the seas. In the name of historical progress—indeed, the march toward the end of history as predicted by Karl Marx—the group of nations thus forming was a whole whose cohesion and durability were guaranteed by the Soviet Union.

Not everything, however, was consistent with this apparent coherence, which the leaders of the USSR presented as an inexorable necessity dictated by the historical laws governing the destiny of people and things (*Zakonomernost'*). Despite these historical laws, however, the nations of the Soviet empire sometimes failed to behave the way they were supposed to. Starting in 1953, from Berlin to Budapest, from Warsaw to Prague, people rose up against the empire and demanded the right to decide their own destinies. Reduced to silence, crushed by Soviet tanks in the name of the laws of history, the overpowered peoples never became resigned. And even in the Soviet Union itself—despite sixty-six years in which the

Soviet people were subjected to terror, to the erasure of memory, to confinement within a total and totalitarian system of values—there were portents of rejection. An unbalanced ethnic demography gave rise to differences and conflicts, the quest for cultural identity, the quest for an authentic history, demands for national languages, and religious revivals. In particular, Islam, when it became a decisive element in the public life of neighboring Iran, was echoed in that part of the Soviet people who suddenly also claimed to be a Muslim people.

These sudden crises or muffled shocks were misunderstood both within and without the USSR. The West, certain that Soviet power was immutable, assumed that no change could come about in the order guaranteed by the USSR. In that state, communism had irreversibly won the day, and all national groups, even those in rebellion, had to adapt to it. After all, wasn't the world moving toward communities of interest larger than the nation-state?

The leaders of the USSR, lulled by five-year plans and grandiose balance sheets, were equally persuaded of the unimportance of these crises. Accustomed to brushing aside—from both themselves and the societies they were responsible for—any news contradicting their march into the radiant future, confident of the ever-growing power of the Soviet state, these leaders listed all problems under the comforting heading of "minor incidents" that could neither affect the course of history nor cast doubt on the changes already achieved.

1986: Andrei Amalrik was off by only two years. It took only an instant for the powerful USSR to discover that its power was a myth and that the continual successes on which it prided itself concealed a general bankruptcy. It has been thought that the source of this revelation lay in one man—Mikhail Gorbachev. In reality, there was also a precipitating event—Chernobyl.

On coming to power in 1985, Gorbachev knew that his country's power concealed many weaknesses. But he also thought that the balance sheet included two undeniable successes—the two empires, foreign and domestic. His goal was to rebuild this power and speed up progress by relying on these two strong points—the Soviet people and the part of Europe economically and militarily integrated into the Soviet community.

Destroying all certainties, the Chernobyl explosion of April 26, 1986, cut Soviet history in two. From then on, there was before Chernobyl and after Chernobyl.

The explosion, which could not be kept secret for long, forced Gorbachev into a radical political turnabout. It was the end of the lie. The Soviet political system could not lie about Chernobyl. Although it issued only partial truths about the incident, it could never again totally lie and impose its own truth on the people. With Chernobyl, the constituent groups of Soviet society suddenly discovered that in the USSR, power, progress, and control over technology and nature concealed only weakness, backwardness, technological underdevelopment, and the destruction of nature. Once they no longer believed everything they were told, the peoples of the USSR rejected everything—first of all, the identity of a Soviet people that had been imposed on them. They demanded that they choose their own destiny. They were unconcerned that by condemning the empire they were also condemning Gorbachev's attempts to democratize it and restore the economy to health.

The Russian empire collapsed in 1917 only to be reincarnated as the Soviet empire. What does this new collapse of an empire mean? Is this only a momentary eclipse that will lead to its reappearance once again in a new form? History is replete with downfalls and resurrections: Alexander's empire gave rise to Byzantium, which later was transmogrified into the Ottoman empire. Or has the empire truly ended, and will modern nation-states or as yet unheard of configurations soon emerge? How responsible for the empire's collapse was the powerful, omnipresent political elite that won power and then held onto it? Was that elite following a plan, or was it simply tossed along by events? Was the Communist party—which in the USSR counted some 20 million members, nearly a tenth of the working population—also expelled from the space where the future of the Soviet people was being played out? Or could it still regain the initiative?

The history of this collapse—and of the relations between this elite, whose power was identical with that of the Soviet Union, and the individuals and ethnic populations whose national projects are proliferating in the USSR—may illuminate this simple question: after the age of revolutions, with the end of the empire, is this the age of nations?

Part One

Incomprehension

1

What Perestroika for the Empire?

PROMOTED to the leadership of the Communist party—and the USSR—on March 11, 1985, Mikhail Gorbachev was faced with a greater challenge than any of his predecessors. To be in control, it was not enough to reach the summit of the system. He still had to eliminate potentially significant rivals—those who stood for the past and who controlled their own fiefdoms. When Gorbachev reached the summit, this power struggle was just beginning.

In the USSR, this struggle took place within the ranks of the Communist party (the "directing core," according to the Constitution of 1977), which distributed jobs and responsibilities. What would be more appropriate for changing the hierarchy than a party congress that met every five years, when it could renew the whole leadership apparatus? All the leaders of the USSR, from Stalin to Brezhnev, used this device, employing various means to eliminate those who inconvenienced them and forging faithful teams who strengthened their control. Gorbachev's exceptional opportunity was that he gained power in March 1985 and was obligated to convene the congress eleven months later, while Khrushchev, for example, had to wait for three years before he could drive out the Stalinists and start abolishing Stalinism. Like Khrushchev, Gorbachev presented himself as a force for change; he claimed to be driving out the Brezhnevists and putting an end to the Brezhnevian stagnation. His watchword was *uskorenie,* or acceleration. A man in a hurry, he wished to speed up change and progress and return to a power that he thought threatened. Destiny provided this man in a hurry an opportunity to get to work quickly.

In Decline, Imperial Success

The Twenty-seventh Congress of the Communist Party, the first congress held under Gorbachev, was full of symbolism. By opening on February 25, 1986, it alluded to the other major congress, the Twentieth, opened by Nikita Khrushchev on February 25, 1956 and devoted to destalinization. After the Twentieth Congress, the USSR was to be different from what it had been under Stalin. Had Gorbachev effected a similar revolution?

Certainly, his report on the state of the USSR, its failures and uncertainties, completely broke with the usual self-satisfied list of nonexistent achievements that Soviet citizens were supposed to believe was their country's reality. In fact, Gorbachev's tone of relative candor and lucidity was well suited to the country's state of intellectual development.

For years the borders of the USSR had not been hermetically sealed, and people had begun to travel. Information coming from the outside about new technology presented Soviets with a different and compelling image of their country. People knew in their hearts that Gorbachev was speaking the truth, but it was not yet a common truth around which society could unite. Beyond the content of Gorbachev's speech, what was revolutionary in 1986 was the switch from individual knowledge to collective knowledge of reality. A civil society worthy of the name cannot be born without this common awareness.

But the USSR is also and above all a multiethnic society. Gorbachev, curiously, had nothing to say about this society's ethnic differences and frustrations. When the empire was mentioned—and he devoted a long passage to it—the otherwise candid Gorbachev reverted to his old demons, and celebratory rhetoric, forgotten for a moment, reclaimed its rights.[1] "The Soviet people represent a new kind of social and international community" in which "oppression and inequalities have been eliminated" and replaced by the "friendship of peoples, respect for national cultures, and national dignity for all," proclaimed Gorbachev, suddenly falling back on the phraseology and tone used by every leader since

1922 to describe the empire. No doubt the USSR was doing badly, but the empire was bearing up well, even if it did have a few marginal problems and shortcomings. In his assessment of history, Gorbachev unhesitatingly referred, under the heading of successes and hope, to what was original and lasting in Lenin's achievement—the integration of the peoples in a state that gave the highest priority to this integration. Because this was the healthy part of the Soviet heritage, Gorbachev concluded that the work of reconstruction could rely on the empire and find in it reasons for hope.

Nevertheless, a closer examination of this passage—whose devotion to an empire, its ideology, and its indefinitely repeated stereotypes refers to a past that Gorbachev otherwise rejected—reveals a seemingly innocuous sentence containing a centralist vision of the national problem. Mentioning economic hardships and the related corruption of the politicians and bureaucrats, Gorbachev specifically targeted what was happening in the peripheral territories. He emphasized that certain republics were "parasites," obsessed by selfish interests and convinced that the Soviet Union existed to support them without their having to contribute to the common good in return.[2] The route to be taken was the opposite of these harmful actions; it was each republic's duty to participate in the development of a single economic complex in which the general interest prevailed over the special interests of the nations comprising the USSR.

Aside from the somewhat contradictory nature of this speech, Gorbachev's vision was quite unambiguous: the USSR had to be a unified economic space. Realizing this goal required a restrictive conception of national cultural rights and cadre policy. Through the use of these two terms, the Soviet regime had long affirmed its respect for national aspirations through the development of all the national cultures and the advancement of national leaders everywhere. Such were Stalin's slogans, even though they were belied by reality, but Gorbachev had no affection for some of the actions taken by citizens from the periphery. Certainly, he said, all ethnic cultures are important, but respect for them must not overshadow their indispensable alliance and their integration into the common ideology, the socialist way of life, and the socialist worldview. There should be no narrow nationalism, but rather national loyalties that would fit into a wider society, held together by the unifying force of socialism.

Gorbachev also denounced the narrow nationalism and localism (*mestnichestvo*) of the cadres, which was of paramount importance at this time for the working of the state and the economy. He said that the political leadership of the republics of the USSR tended to withdraw into themselves, to practice a policy of "national favoritism," to prefer any fellow countryman to a competent Soviet when it came to promoting cadres. Gorbachev noted that this attitude led to discrimination against all those who were not members of the main nationality of each republic and who were denied the positions to which they were entitled by their competence or the size of the group they belonged to. Gorbachev was accusing the republics and the great national regions of the USSR of practicing "national preference" or nepotism—the corruption inherent in any policy that favored the group and of thereby imperiling the general economic interest and the progress of interethnic relations.

A Contrary Trend Toward Local Actions

On the one hand, in 1986 Gorbachev presented an idyllic picture of the national problem that, in light of the rest of the report, seemed to represent the only real success in all Soviet history, a success on which to base a policy of recovery. On the other hand, he seemed to be unreceptive to the demands for action from the peripheral republics, which, he suggested, had taken too much advantage of the Brezhnevian stagnation by looking after their own interests. Unlike Brezhnev in his final years in power, Gorbachev did not say that it was time for the republics to start "repaying their debts to Russia." Nevertheless, behind the rhetoric of traditional satisfaction appeared a certain irritation with the overdemanding nationalities who refused to play the game of common development.

Speaking of a "Soviet people" as an incontestable reality and yet attacking the parasitism and localism of some of its elements was only apparently contradictory. Gorbachev's remarks clearly suggest that, for him, it was progress and the very existence of a Soviet people born of so many different peoples that were decisive, and that the regrettable actions he denounced resulted partly from "relics" of a past that was close to disappearing and partly from a policy of stagnation that had distorted Soviet development.

Despite Gorbachev's sometimes caustic criticisms, underlying this judgment is an optimistic view of the work accomplished in this area by his predecessors. In the pessimistic evaluation of his country that he presented to the congress, Gorbachev held on to the national question as the chief element of comfort and support for the future. All the same, his criticism led to proposals to give perestroika some concrete content. Attacking the "parasitism" of certain republics, he immediately found in this an argument exhorting them to exceptional mobilization in the economic effort to which he was committing society. Parasitic and incompetent, these republics could make a considerable effort toward the common good they had so long neglected. National preference, nepotism, and corruption—these accusations led directly to a purge of the local cadres and to the possibility of replacing them with people whose competence and utility would be the real criteria for selection. In this respect Gorbachev held a marvelous instrument for breaking up the strongholds of power that Brezhnev had allowed to be created and consolidated and that could only make difficulties for the new secretary-general. Even more, by attacking the principle of localism in the selection of cadres, Gorbachev made it possible to send the people he wished to replace them and hence to form his own teams throughout Soviet territory. Deploring the small number of "minority national groups" (especially Russians and Ukrainians) in leadership posts in these republics, he thus opened the way to a reshuffling of their leaders along less national lines and thus to eventual progress in internationalism, another key term in the speech.

The nail first put in place by Gorbachev was then hammered in by Yegor Ligachev, who appeared at this congress as the number-two man in the party. He openly advocated an "exchange of cadres from one republic to another, from the center to the periphery, and from the periphery to the center."[3] An exchange of cadres meant the "parachuting" of leaders, an old policy of the Soviet government intended to prevent the buildup of stable national elites that would not legitimately represent the people. Since the early 1920s, the government had pursued a twofold goal. One was to nurture leaders necessary for the country's development. These had to be chosen from all ethnic groups in order to avoid imbalances between the dominating nations and the dominated nations, which could produce frustrations and rebellions. The other goal was to avoid the formation of leadership elites rooted in their own nations and cultures of origin and instead create a large Soviet elite, the vanguard of the Soviet people to come. The Stalinist compromise between national culture and proletariat culture was most concretely expressed in these two goals. The national societies provisionally retained their own cultures, but the Soviet leadership elite was the embodiment of the proletarian culture that would someday be common to all. Khrushchev and then Brezhnev made concessions to the national societies that were eager to appropriate their elites; but both leaders worried about the effects of a possible domestic alliance between the people and its elites. And Brezhnev, at the Twenty-sixth Congress meeting a year before his death, raised a cry of alarm, noting that at the periphery, the Russians, the human glue of the system, were in a position of weakness relative to the more self-confident national elites.[4]

In his 1986 discussion of the national problem, Mikhail Gorbachev plainly agreed with the party and with the reactions of frustration expressed by the Russians. In national matters, the new program adopted by the Twenty-seventh Congress was little different from the preceding one. Its major innovation was its use of the concept of the "united Soviet people" (*edinyi sovetskii narod*), previously absent from this kind of statement, even if Soviet ideologists had used it widely since 1977. When the program was published, *Pravda* opened its columns to public debate.[5] In assessing public reaction, one should keep in mind the insistence on Russia's role in Soviet history—on the Russian contributions to raising other federated peoples from underdevelopment, on the need for further development in the common use of the Russian language and its civilizing role, and on the need to draw more clearly the prospects for a future unitary state for a unified people.[6] No doubt *Pravda* chose to emphasise the frustrations of the Russians and the conclusion to be drawn from the certainty that the Soviet people had become a reality. But this choice reveals some existing orientations within the Communist party apparatus, which in 1986 was still the dominant political organ in the Soviet system. It is significant here that, contrary to his three immediate predecessors, Gorbachev did not at the time appear to be striving for any title other than that of the party's secretary-general, leaving to others the job of embodying the state (Mikoyan) or the government (Nicolai Ryzhkov). The line defined by the party at its Twenty-seventh Congress was thus very much the one that always had controlled all the orientations of the USSR.

Reconstructing the Empire

In December 1986, fighting broke out between Kazakhs and the police in the city of Alma-Ata. (I shall return to this event later.) These first riots in Gorbachev's USSR did not, however, appear to shake the secretary-general's confidence nor that of the party leadership. Addressing the Central Committee meeting in a plenary session a few weeks later on January 27, 1987,[7] Gorbachev quickly tackled the problem of the conflicts arising in a national setting, but he did so in a precise and candid way that departed from tradition. For the first time, glasnost seemed to approach areas where optimistic ready-made formulas previously had prevailed. Although he admitted that the USSR's multiethnicity could present problems, Gorbachev did not suggest any revision of the nationalities policy. The causes of the clashes were ones he mentioned in general terms to the congress: localism, ethnic isolationism, even "national arrogance" (that is, behavior inherited from the past that the Brezhnevian stagnation had not eliminated). Like his predecessors, Gorbachev thought that the remedy for these relics lay in an intransigent fidelity to the internationalist ideology and in an increased "internationalization" of the cadres, that is, the abandonment by the nationalities of their discriminatory attitude toward the other groups. These familiar proposals did not stir up new debate.

Gorbachev, however, was not insensitive to the deterioration of the national mood. The press and the experts began debating the specific difficulties that the central government was encountering at the periphery. The notion of ethnic conflict even made an appearance.[8] It is scarcely surprising that, celebrating the seventieth anniversary of the October Revolution, Gorbachev did not dwell on this problem. His remarks revealed his constant discomfort with the national question, even an inability to grasp the facts.

Gorbachev once again repeated that the Soviet regime could resolve the national problem, but he conceded that relations between the nations continued to be complex and called for the party's constant vigilance. For the first time since his assumption of power, he included this question as part of perestroika and democratization, not in the form of a promise to rework the nationalities policy but as an element to be introduced in the general debate. This was a small but significant shift of emphasis.

It was small in that he still thought—and said—that the friendship of the people of the USSR—that myth that runs throughout Soviet history—was a success and must not be undercut by experimentation; and because he was still convinced that the power of the Soviet state and its historical glory were linked to this success. These certainties led him to condemn in advance anything that might produce instability in this area. But the shift was also significant in that he accepted that here too the taboos were evaporating and that glasnost was applicable. By committing himself in November 1987 to a debate about problems connected with a multiethnic state, Gorbachev implicitly noted that these problems were more serious than he had acknowledged, and he accepted the idea that these problems, like all others, should be debated publicly.

Actually, he was only following closely on the debates and events. Since the summer of 1986, he had begun losing control of glasnost, and the limits he assigned to it were not observed, notably in the peripheral regions. In recognizing the legitimacy of this debate, Gorbachev seemed to be reaching out to those who had held it. Was he going to change from an intransigent orthodoxy to a recognition of national interests, as Khrushchev had done in 1956? Even though only momentary, this change had had enduring consequences.

Russians Come to the Aid of the USSR

Frequently contradictory in his remarks about the national problem, Gorbachev was much less so when the central government and its leaders were at issue. Seemingly assured of Soviet solidity and control, he paid scant attention to national sensitivities, and he blithely overlooked the rules for national representation that had been in effect since 1956.

By placing new people in institutions of power—notably the Politburo and the party secretariat—Gorbachev showed a strong bent for centralization.[9] Two signs made this perfectly clear: the Russianization of the system's highest leadership cadres and particularly their ignorance of events occurring on the periphery.

The first feature is shown by a comparison of Brezhnev's final Politburo and the one organized in 1985 and 1986. True, the often-deceptive figures suggest a certain stability in interethnic relations. In 1982, the Politburo had three non-Russian members out of thirteen voting members. The same was true in 1987. But the two trios were very different in their national representativeness. Leonid Brezhnev's colleagues were close to him because, with one or two exceptions, they were the highest leaders of their republics. But the Ukrainian Shcherbitsky, the Kazakh Kunaev, the Azeri Geidar Aliev, the Georgian Shevardnadze, the Uzbek Rashidov, and the Belorussian Kiselev were all first secretaries of the Communist party and came to the Politburo to defend the interests of their own republics as much as to set a common policy to be imposed on their compatriots.

Two years after Gorbachev reached power, things were quite different. Although Shcherbitsky remained in the Politburo representing the Ukrainian party, the two other national representatives no longer had any authoritative tie to their native republics. Until 1985, Shevardnadze was the first secretary of the Georgian Communist party and was then promoted to minister of foreign affairs, putting him in the highest category of the full members of the Politburo; but his party was no longer represented in it. The same is true of the Belorussian Sliunkov, who entered the Politburo to replace a compatriot (the former first secretary of the Belorussian party, Kiselev) and, promoted in 1987 to secretary of the Central Committee, from then on had a seat in the Politburo. As for the six deputies of the Politburo, they were all Russian.[10]

Not only did national dignitaries rarely represent their countries of origin, but more seriously yet, whole areas of the periphery no longer sent contingents to the party's higher authorities. Under Brezhnev, full members and deputies spoke for of the Muslim republics of Central Asia and for the Caucasus, Georgia, and the two Slavic states, the Ukraine and Belorussia. Some, such as the Baltic peoples and the Armenians, were certainly absent, but at least the most numerous peoples were represented. In 1987, all the Muslim republics and the Caucasus disappeared from the Politburo without any of the missing ones replacing them. Russian predominance was coupled with a near-omnipresence of Slavs. And although in the secretariat of the Central Committee Nikolai Sliunkov was not Russian, he was still a Slav.

It should be recalled that at certain times the secretariat had opened its doors to several secretaries from outside the Russian majority. In 1960 there were three of them, one of whom was Mukhitdinov, a Muslim who previously headed the Communist party in Uzbekistan. The abrupt exclusion of the southern part of the USSR—the most heterogeneous in culture and size of population, the most dynamic demographically, and generally the poorest—in favor of Slavs, who were connected through culture, religion, level of development, and social behavior, from then on constituted a serious problem. A relatively homogeneous world therefore figured in the decisions that were made at the time of perestroika.

In addition, in contrast to earlier practices, the Russians at the center of the decision-making system knew only Russia. For decades, high Soviet leaders exercised their skills in various republics before arriving in Moscow and generally acquired from these travels a fairly complete view of problems at the periphery. With Mikhail Gorbachev, everything changed. Only two men in his immediate entourage had worked in a national setting: the head of the KGB in 1987, Viktor Chebrikov, who at one time worked in the Ukraine, and General Yazov, the minister of defense, who for a short time commanded the military district of Central Asia. A military command, however, is not likely to inspire mixing with the population, but rather the contrary. And being in the KGB does little to promote one's popularity. It is doubtful whether the Ukrainians or Central Asians recognized the representatives of the government as authentic spokesmen for their aspirations. Other than these two cases, no member of the Politburo and no secretary had had any experience in the border countries of the USSR.

This situation, new in the post-Stalinist era, had two results. For the nationalities, a sense of being ignored and even scorned by Moscow led, when conflicts arose, to the certitude that problems could be dealt with effectively only on their home turf—and by force. With no representatives in Moscow, how could it be otherwise? Around Gorbachev in Moscow, however, the effects of this Russianization and Slavicization of the government's top personnel were no less damaging. Lack of experience led the whole Soviet leadership to ignore mounting pressures at the periphery and, later, to explain it by the familiar general causes: corruption, hooliganism, defects in the organization of the cadres. It also led them to underestimate each event that disturbed the periphery and the series of crises. Gorbachev and his top aides could barely tell the difference between an interethnic conflict and a riot in front of an empty store. They were largely oblivious to the details of the problems posed by the nations.

The source of this blindness—of such an error—for federalism presupposes the participation of all constituent groups in the decisions made by the community—is surely the problem of one man, Gorbachev. Everything that made him attractive to the world—his modernity, his advanced education compared to that of his predecessors, his "European" ways—had another side: he was Russian, born within the borders of the empire, but had come early in life to Moscow to get his education, and had no experience of life and work outside the Russian (and hence European) setting. He represented the developed as against the underdeveloped USSR, the USSR oriented toward the West in contrast to the one that wanted to get back to its roots, the USSR with ties to Christianity as opposed to Islam. This fortunate man had traveled for weeks in France during his youth and had surely learned a good deal about it, but he had not had a similar experience of his own country. He also did not know people from the periphery and their distant cultures on whom he could someday rely. Stalin had a good knowledge of the Caucasus and the Caucasians; Khrushchev, of the Ukraine; Brezhnev, Moldavia, and Kazakhstan; Andropov, of the Karelian Republic. Even the most insignificant of general-secretaries, the ephemeral Konstantin Chernenko, had once held a high post in Moldavia. All had acquired knowledge and formed friendships in these republics. Gorbachev had done so only in Russia. Certainly, he had confidence in a Georgian to lead the USSR's foreign policy, but this Georgian, who had long been a high official in the KGB and was well acquainted with his compatriots and their aspirations, was, because of this flattering appointment, diverted from the domestic problem where his experience could be of value. Consider the list of Gorbachev's advisers, who for the most part were Russian. Although the economist Aganbegian was Armenian, he was an Armenian from Moscow who was completely assimilated, and who rediscovered his origins only when two disasters, massacre and earthquake, overtook his own people. Should we be surprised that, for those who were not Russian-born, Gorbachev was perceived mainly as a Russian?

The misunderstanding between Gorbachev and the nationalities of the USSR began as early as 1985. The man who passionately wanted to be Lenin's heir thought of the USSR in Leninist terms. For Gorbachev, the important thing was neither the nation nor federalism but the Soviet whole that he wanted to rebuild. Like Lenin, the only part of the USSR Gorbachev knew was Russia; like Lenin, he thought that this was of small importance, for power, concentrated at the center of the country, could resolve everything. To Stalin's beloved slogan, *kadry vse reshaiut* (cadres solve everything), Gorbachev could contrast a slogan that Lenin had inscribed with the facts and to which he gave renewed life, *vlast' vse reshaiet* (power solves everything), and power knows no national or territorial borders. Convinced that, armed with power and an irrefutable plan, he would be understood, Gorbachev did not see that his fierce Leninism could only stimulate rebellion in the already ruinous climate of the USSR in the 1980s. The weeding out of corrupt leadership cadres, which was indispensable for a policy of rebuilding, was to heighten the misunderstanding and hasten the process of disintegration to which the new secretary-general had long remained oblivious.

2

The Mafiocracy

CORRUPTION was not unknown in the USSR. The political system that sang the praises of egalitarianism, as it obscured the privileges of the *nomenklatura,* encouraged abuses, trafficking, and corruption. But the ideology that constantly asserted that communism had produced a "new man"—profoundly honest and devoted to the common good—blamed criminal activities on those at the fringes of society. Deviant behavior was traced to abnormality and was said to call for psychiatric rather than punitive treatment. Surely those who had been in the concentration camps and borne witness—like Evgenia Ginzburg, Shalamov, Alexandr Solzhenitsyn, Bukovsky, and many others—had reported a chilling picture of Soviet criminality. Anyone who read *The Gulag Archipelago*[1] could not have the slightest illusion about the "new man" and knew that corruption in the USSR matched that in other societies. But the Soviet press had long projected the image of moral behavior that the system demanded. Crime reports were extremely rare, and letters to the editors of newspapers that occasionally tried to break the silence were carefully screened out. After all, criminality could not be reconciled with the articles and images systematically glorifying the deeds of the Soviet workers.

Inevitably a scandalous episode erupted in this immaculate universe and called into question its veracity. In 1972, when Eduard Shevardnadze succeeded Vasili Mjavanadze as head of the Georgian Communist party, there was no doubt in the USSR of the real reason for this change—corruption. Mjavanadze cultivated a system of corruption that spread throughout the political life of Georgia. Everything was an occasion for payoffs, and the party's first secretary and his wife surprisingly prefigured Ceausescu and his wife: ambitious, arrogant, wielding their authority in every aspect of public life, they acquired dachas, furs, and diamonds. Even at the time, it was clear that money had little meaning in the USSR, so that the only things that counted were goods. Although Mjavanadze fell from power, the government remained silent about the corruption that had caused his disgrace. Shevardnadze was given responsibility for cleaning out the Augean stables, and little was made of corruption that had reached extraordinary proportions.

A short time later, in 1976, the first secretary of the Ukrainian Communist party, Shcherbitsky, suggested that his predecessor, Piotr Shelest, had lost all sense of morality. Once again, however, the desire to maintain an image of an ideal Soviet society undermined or contradicted efforts to uncover corrupt practices that the people suspected without realizing their extent.

In 1979, the "caviar scandal" broke. It was learned that caviar was hidden in cans labeled "herring," and sold to luxury stores or sent abroad at the price of caviar, the price difference between herring and caviar being pocketed by the network's organizers. The scandal exposed the reality of a criminal organization that compromised the USSR's minister and vice minister for fishing, the secretary of the *obkom* of Krasnodar, the president of the Municipal Council of Sochi, and many other politicians. But since most had connections with Brezhnev or Kosygin, and since many top civil servants were also implicated in the affair, the central government opted for a certain indulgence and discretion. Soviet citizens knew something about the matter, but they learned about it not through their newspapers but through radio broadcasts from abroad that raised an uproar about what deserved to be a real affair of state.[2]

The silence held sway until 1982 and the death of Brezhnev. The corruption around the elderly secretary-general—his greed for profitable decorations, the royalties from mediocre books he had not written, the foreign luxury cars he collected—started in his own family and infested everything. At the time, however, official corruption was so insolently flaunted that the people, aware of the system's tendencies, considered it normal to behave similarly. This ubiquity of corruption led Andropov, on his accession to power, to suddenly and often openly react.

On December 11, 1982, the press, with *Pravda* at the forefront, reported that a Politburo meeting had been devoted to discussing corruption and condemning this moral degradation within the governing class.[3] On December 18, a decree effective on January 1 instituted a system of sanctions designed to halt corruption. This was described in detail by the press, which had finally been freed from its obligation to remain silent. Andropov's campaign, even though slowed by his illness and then by Chernenko's coming to power, could not be stopped.

The scandal even extended to Brezhnev's family; it became impossible to revert to silence. In the two years between the start of the crisis and Gorbachev's rise to power, quite a few members of the *nomenklatura* disappeared from the scene—forced into disgraced retirements or suicides that were declared to be heart attacks. The forced resignations, carried out as part of the struggle against corruption, so strongly affected the public that every sudden death at the system's center was immediately linked to corruption. Each death was suspected of cloaking a suicide—a good way to escape justice. Characteristic of those two years were a creeping obsession with corruption and with its existence at the center of the system—that is, in Russia itself. Until 1985, attention on the embezzlements and misappropriations of funds among the high-ranking *nomenklatura* seemed clearly to imply that a maximum of power supplied a maximum number of ways to profit from it.[4]

From Corruption to the Mafia: The Uzbek Question

With the Rashidov affair, the corruption problem suddenly switched from the central government to the periphery and took on an entirely new dimension.

Under the Brezhnevian stagnation, Sharaf Rashidov had been one of the central government's Muslim darlings. He was made head of the Communist party of Uzbekistan under Khrushchev in 1959, had presided since 1962 over the destinies of his republic, was a deputy minister of the Central Committee, and had been blanketed with honors by Brezhnev. He was an ideal subordinate for Brezhnev, for he forcefully supported all his master's plans and demands, no matter what lies he had to tell. On two decisive matters, Rashidov had long convinced the central government that Uzbekistan—by 1979, the USSR's fourth-largest republic (presently the third-largest) and the largest Muslim one—was a model of economic and cultural integration. In the early 1980s, he had planned and announced unusually large cotton harvests (over 13 million tons a year; Uzbekistan is the USSR's largest cotton producer), which was bound to satisfy Brezhnev's megalomania. "Ever more!" he announced, and Moscow was pleased with these unlikely production figures.

The same was true of another theme dear to the heart of the team in power, the headway made by the Russian language among non-Russians. Rashidov made public the incredible successes of his linguistic policy from one census to another.[5] In a few years, it appeared that the whole population of Uzbekistan, including the elderly and infants, had succeeded in becoming perfectly bilingual. If the Soviet people, speaking a common language and totally devoted to the common good through economic effort, existed anywhere, it was in Rashidov's Uzbekistan! Hence, it was not surprising that Brezhnev pinned many a decoration on him nor that, shortly after Brezhnev's sudden death in October 1983, the Soviet government rewarded him with honors due a member of the Politburo.[6]

Unfortunately for Rashidov's memory, the following year Andropov's assault on corruption reached Uzbekistan, where it was murmured that the grandiose economic triumphs that had secured Rashidov's glory were due more to manipulated numbers than to any verifiable reality. The Central Committee of the Uzbekistan Communist party met in a plenary session in June 1984 and echoed these rumors but left the late Rashidov out of any disagreeable mutterings. In the two years from 1984 to 1986, Uzbekistan underwent a systematic purge: many party and state cadres were eliminated, but disgrace did not yet extend to the top of the hierarchy, and the case of Uzbekistan did not seem out of the ordinary.

In 1986, a radical change took place. In January, the Uzbek Communist party congress unhesitatingly attacked Rashidov and his corrupt management techniques. Rashidov was also one of the stars of corruption at the Twenty-seventh Congress of the Communist Party of the Soviet Union held a month later. His posthumous condemnation was spectacular: all the honors rendered him on his death were revoked, and his remains were removed from the official cemetery in Tashkent.[7] With him fell Uzbekistan's former prime minister, Khudaiberdiev, driven to a seemingly honorable resignation (given his age) in November 1984 but expelled from party ranks in 1986. Since Rashidov's death spared him condemnation, his whole entourage, starting in 1986, was dishonored.

This time talk was not to be muffled. In Moscow, there was a denunciation of the huge "Uzbek conspiracy" that led to systematic economic falsification, concealing the realities of a republic riddled with corruption and nepotism. The harm these deceptive practices did to the Soviet economy explains the severity of the subsequent purge. Death sentences were carried out on some local officials, such as Shahrizabz in Bukhara, while others received heavy prison terms and expulsions. The entire hierarchy of the republic was overturned as revelations continued. In Uzbekistan, everything was corrupt: not only had the economy been sabotaged, but the whole system of access to jobs was undermined by criminal practices. The university system, which provided leadership cadres, operated at an extremely low level, for the admission of students was largely a matter of connections, kinship, and payoffs. The teaching staff complied with these arrangements—and owed their diplomas and positions to the same standards. The Soviet press made the traditionally prestigious state-run University of Tashkent an object of ridicule.

The most serious accusation was that in Uzbekistan, the advancement of leadership elites by the bestowal of diplomas had been a function of membership in a national group and not of abilities. Since the late 1950s the Soviet government had made great efforts to develop national elites, granting considerable funds to the educational system. This effort to ensure the cultural alignment of the less-developed—Muslim—periphery with the rest of the USSR was thus undermined. These leadership elites were phony elites that reflected no progress. In a few months, Uzbekistan plummeted from the rank of vanguard republic to that of a country that was intellectually, morally, and economically underdeveloped.

Several elements of this general denunciation are worth emphasizing. Corruption in Uzbekistan was presented as a phenomenon that had occurred on a scale quite different from that in the rest of the USSR. It soon appeared that this phenomenon was so widespread and deep-rooted that its solution no longer lay with the Uzbeks. More thoroughly corrupted than other Soviet citizens, they were unable to rid themselves of their unethical habits. As time passed, more revelations uncovered the magnitude of the Uzbek crisis and the difficulty of ending it. The central press[8] eagerly observed that it was not enough to dismiss hundreds of officials and thousands of employees. Corruption in Uzbekistan was gradually isolated from the general problem of corruption and became the Uzbek question—a condemnation of the whole people and their political tradition.

The indexing of Uzbekistan was further aggravated by the dissemination of data on the incidence of ordinary criminality in the republic. Here, a decisive role was played by the weekly news magazine *Ogoniok*, the most ardent defender of Gorbachev's plan of restructuring. It conducted a systematic investigation that led to a disturbing revelation, one that was overwhelming for a USSR that only recently had removed its moral blinders: in the extent of its organized crime, the USSR kept pace with the most degenerate societies.[9] And leading this previously unsuspected criminality was the Uzbekistan mafia, a network of well-organized and specialized gangs as large and influential as the Sicilian and American mafias.

Not only did the *Ogoniok* investigation disclose the existence of these networks and provide a detailed description of their organization and activities, but it also dwelt on the link between this mafia and the republic's political and administrative elite. In fact, *Ogoniok* concluded that the mafia pervaded the whole of Uzbek life, controlling both economic corruption and ordinary criminality. It ran a protection racket—that is, the racketeering of both legal and illegal individual enterprises, monopolized drug trafficking, and controlled prostitution and murder for hire. The price for getting rid of some objectionable person was even codified. For 33,000 rubles, someone could hire a killer provided by the crime syndicate to eliminate a political rival, a bothersome husband, even a *sovkhoz* director who would not close his eyes to the illegal activities of his accountant. In the ensuing trial, the Uzbek mafia's arrangements with government and party leaders (including police chiefs), its recruiting of elite university graduates were regarded as perhaps less serious than the revelation that it was organized along lines coinciding with the old social structure of clans. Those dealing with the mafia suggested that this was no superficial phenomenon, but a continuation of ancient traditions in Uzbekistan.

The affair caused a considerable uproar. The press gave the facts a full airing. At the Party's Nineteenth Congress in July 1988, the Uzbek delegates were repeatedly attacked for their complicity with the mafia. What did the people of the Soviet Union, sitting in front of their televisions, think of a nation-republic that was represented on this ceremonial occasion by notorious criminals? By tracking down the Uzbek mafia, two judges, Gdlian and Ivanov, gained the reputation of incorruptible defenders of the law and became highly popular.[10]

There was no longer any doubt that Uzbekistan was dominated by corruption and criminality. Glasnost, however, taught the Soviets that their entire country had been systematically bled by both the *nomenklatura* and the gangs. Was Uzbekistan an extreme case? Perhaps. But by highlighting it, the central government provoked sharp and irreparable reactions among the Uzbeks. Attacked so forcefully, they felt that they were being systematically discriminated against. Some Uzbek intellectuals, who were otherwise integrated into the USSR, interpreted these reprimands as humiliation. In a rousing article,[11] the journalist Kamal Ikramov, the son of Uzbek party head Akmal Ikramov, whom Stalin had liquidated in 1938, waxed indignant: did the pretext of purifying a corrupt elite necessitate the creation of an Uzbek question, as though a whole people were collectively responsible and collectively criminal? This Uzbek "distinctiveness" ultimately led the people to rally around their leaders, as questionable as they were, in opposition to the larger country that condemned them.

The questioning of the competence of the elites added to the people's humiliation and anxieties. If Uzbek elites were denounced for incompetence, would it be logical to place the republic under the authority of more competent elites—that is, the Russians? The need for the "movement of cadres" that Yegor Ligachev praised at the Twenty-seventh Congress was given a new justification. Uzbeks suspected that they were being attacked for being "worse" than the other republics because "Russianization," operating in the party organs since Gorbachev had reached power, deprived them of high-level defenders who could reestablish some balance in the analysis of the phenomena of corruption.

Another possible reason was that Russianization at the top was not fortuitous: Russians, who represented a decreasing proportion of the Soviet population, seemed intent on maintaining their authority over the whole in the name of some moral superiority. Stalin had once proclaimed that the Russians had been the country's ablest defenders during the war. Denouncing "collaborationist peoples,"[12] he organized the nations of the USSR into a hierarchy and legitimated the claim of the Russian people as "the best" to guide the others.

In 1987, the Uzbeks asked if the hierarchizing of corruption and crime didn't have the same function—to give the Russian center, where the "purge" operation had started, a new legitimacy in guiding all those who had fallen into crime, people who were deeply corrupt and criminal. Other peoples once had been accused of collaboration and relegated to the bottom rung on the scale of patriotism for the same purpose. This growing suspicion among the Uzbeks fueled not only national but also Islamic solidarities. It is not irrelevant to observe that Ikramov's outrage about the fate to which the Uzbeks were exposed was reported on by the Azeri writer Mirza Ibragimov.[13] Thus the conviction grew that the struggle against corruption served numerous purposes, that it helped in the final analysis to establish Russian authority over Muslim people, whose demographic growth and economic problems should, on the contrary, benefit from a more just distribution of responsibilities and resources. At the Muslim periphery, the rejection of a purge and, beyond that, of a perestroika initiated by Russia, which would be the main political beneficiary, was thus one of the consequences of the campaign against Uzbekistan.

The Eternal Villain

The hierarchy of criminality set up by the purge put Georgia in a category close to Uzbekistan—that of incorrigibles—with the same results: in 1972 it was the first republic to suffer a purge for "corrupt doings." Because this purge lasted thirteen years, until Gorbachev came to power, it might have been thought that Georgia would be spared during the great cleanup. The Georgians had already been surprised by Shevardnadze's constant and numerous accusations and dismissals, and they insinuated that what he was so obstinately seeking to eradicate in their republic was not corruption, but the sense of national pride. In 1985, when he was called to succeed Andrei Gromyko as minister for foreign affairs, his successor Dzhumbar Patiashvili in turn pursued a purge campaign against the cadres promoted by Shevardnadze, who was called "Mister Clean." In a few months, he eliminated three ministers, two secretaries of the Central Committee, the secretary of a regional party committee, several high officials in the Communist party and the government, two heads of newspapers, and top officials in the police. Without exception, all were accused of corruption and nepotism.

Even though he quoted the authority of his predecessor, Patiashvili did not hesitate to let it be known that the reign of Shevardnadze, the purifier, had brought about no change in the republic's moral climate and that everything there had to be redone from square one. At the Twenty-seventh Congress, he maintained that the most conservative tendencies in Georgia still survived; that the republic was dominated by a general desire for acquisitions that, at all levels, fueled a morality contrary to that of socialism. It perpetuated deviant behavior, such as corruption, economic sabotage, and the proliferation of clandestine activities that undermined the legal economy. Patiashvili said that Georgia could purge this mentality oriented to private property only by real class warfare.[14] Did this militant speech, with its indefinitely repeated accusations against a corrupt society, mean that Shevardnadze had been ineffective in his fight against corruption, or that Georgia, like Uzbekistan, was by tradition and culture a propitious setting for such tendencies to emerge?

The latter idea was widely accepted in Russia, where affection for Georgians had long ago given way to exasperation with what was called "Georgian racketeering" which meant no more than their efficiency as small business people, legal or otherwise. Over the years, the Georgians had seemed more skilled than other people in profiting from parallel activities that flourished in the USSR and made it possible to improve an arduous daily life. Legends circulated about Georgian wealth and illegal activities that survived Shevardnadze's housecleaning during the 1970s, and which sharply inhibited his compatriots' spirit of initiative. On the whole, Soviet public opinion was prepared to accept the idea that, like the Uzbeks, the Georgians still had to undergo a severe purge.

But in Georgia as in Uzbekistan, this treatment was considered discriminatory and couldn't have had more negative effects. Although in the 1970s, Georgians readily recognized the misdeeds of Mjavanadze and his cronies, they thought that the main purpose of Shevardnadze's purification was to check their independence of action. They were well aware that the practices denounced were not confined to their republic; in fact, it was perfectly clear that these had become ever more widespread during the Brezhnev years. Tending to think as early as 1972 that they were victims of a policy that was more anti-Georgian than moralistic, they gave free rein to their exasperation—notably during the large demonstration of 1978.

Up to 1985, however, this exasperation was moderated by a certain confidence in Shevardnadze's ambiguous political game. Certainly, he purged the republic at Moscow's orders, but he also defended it in Moscow. A member of the Politburo, he publicized the claims of Georgia's national interest, and in the face of demonstrators in Tbilisi in 1978, he had been able to forestall an intervention by tanks that encircled the provincial capital. He was also able to get Moscow to admit that the cultural demands that were the source of the demonstrations (the Georgians wanted the Georgian language registered as the official language in their constitution) should be accepted, if uncontrollable outbreaks of fury were to be avoided. After 1985, Georgia no longer had a representative in Moscow, and to his compatriots, Patiashvili was merely a tool of the central government assigned to decapitate Georgian leadership cadres and reduce a humiliated people to silence. In the frenzy of purification dogging them, the Georgians, like the Uzbeks, saw the policy as a way to crush the most deeply rooted nationalisms and to justify the increased power of the center over the periphery.

Corruption as a Pretext for Russianization: Kazakhstan

Although the Uzbeks and Georgians deserve a special mention among the prizewinners for criminality, the other republics did not emerge unscathed, and all of Central Asia ranked high. Once again, the implicit theme—the denunciation of a link between corruption and social traditions—expanded on the policy of purification bequeathed by Andropov to Gorbachev. All the republics shared in humiliating incidents. In Kirghizia, a brutal purge culminated in the dismissal on November 2, 1985, of the first secretary of the Communist party, Usubaliev, who had held the post for nearly three decades and who, with several of his collaborators, was ignominiously expelled from the Communist Party of the USSR.[15] He was the first high official in Central Asia to undergo this fate. Until then, dismissals were more secretive. In Turkmenistan and in the Caucasus and Azerbaijan, wave after wave of purification also suggested the presence of deep-seated ills could not be overcome by the interested parties themselves. It is the case of Kazakhstan, however, that sheds most light on the nature of the conflict between the periphery and the center concerning the real problem of corruption.

The purge in Kazakhstan is interesting because of its effects on the membership of the Politburo at the center and on interethnic relations at the periphery. In 1985, Kazakhstan carried great weight in the USSR, even though the population increase in Uzbekistan had reduced it from third to fourth place. A population of 14.5 million people, a sensitive geographical position on the border with China, and considerable industrial resources assured this republic of representation in the Politburo. The first secretary of the Kazakh Communist party, Kunaev, one of Brezhnev's close associates, had been a member of the Supreme Soviet since 1971. When corruption became a widespread concern in the Soviet government, Kunaev maintained a low profile. The attack on corrupt practices in his own republic came from Moscow, not the local government, as was then usual.[16]

It was *Pravda* that launched the assault on the corrupt practices and abuses of power that permeated this republic. The campaign thus undertaken shortly led to the expulsion of half the secretaries of the Kazakh Central Committee, several regional party officials, and numerous cadres from all sectors. In the midst of this upheaval, Kunaev at first seemed immovable. His whole entourage, his collaborators, and even his relatives were discredited and brought before the courts. He never defended them and sometimes accused them, while taking part in purification discussions only from afar. At the Kazakh party congress in February 1986, he criticized some leading officials of the republic.[17] There was no sign, however, that he backed the drive for general renewal endorsed by Moscow. Nevertheless, he survived the Twenty-seventh Congress and kept his seat in the Politburo, and only after new attacks, once again from Moscow, did he fall.

As in Uzbekistan, all the actions of the Kazakh administration were denounced. The region's economy had been misrepresented, and its resources were systematically bled (the leaders had had a racing stable built by diverting funds and materials from the *sovkhozes* and managed by incompetents). Here we return to the problem of the local leadership cadres and their capacity for assuming responsibilities. At issue in Kazakhstan was national and even clan favoritism as a criterion for admission to institutions of higher learning.[18] Since education in Central Asia had been thus led astray, it was logical to fill command posts with people competent enough to set things straight—that is, Russians. It was in Kazakhstan that Gorbachev was to draw conclusions about the accusations of nepotism in Central Asia, thereby justifying all the suspicions surrounding the policy of purification and provoking the first violent clash between ethnic groups.

Kunaev was dismissed on December 16, 1986; he lost both his party leadership and his seat in the Politburo. Although the dismissal of an elderly (over seventy) person who was a close companion of Brezhnev and whose entourage already had been eliminated was predictable, the selection of his successor was less so. At first sight, the man named to the office, Gennadi Kolbin, was well suited to the new style of Soviet leadership. Close to Gorbachev in age—fifty-nine—and university education, he had solid credentials as a competent and effective apparatchik. In

Georgia he had worked with Shevardnadze and in Sverdlovsk with Nikolai Ryzhkov, where he battled corruption and alcoholism during the Andropov administration. But most important, Kolbin was a Russian. Since Stalin's death, a tacit agreement had governed the distribution of power between the Russians and non-Russians within the parties of the various republics. To the non-Russians went the post of first secretary, the "boss" of the republic, who worked hand in glove with Moscow and had an official seat in the Central Committee of the Communist party of the Soviet Union and sometimes in the Politburo; to the Russians went the post of second secretary, who controlled all the nominations. This balance, which had been observed since 1956, had sufficiently reassured the various nationalities and convinced them not to dispute the nominations to these two posts. The arrival of the tradition-breaking Kolbin put two Russians at the head of the Kazakh party and underscored a change in policy already suggested by the Russianization of the central governing bodies of the party. Not only was Kolbin Russian, but he had no experience of Kazakhstan or much of the Muslim setting. His nomination indicated a desire to appoint a person who had never shown any affinity or common interest with the citizens of his new republic.

In resorting to the long-abandoned method of "parachuting," Gorbachev was indicating his mistrust of the national leadership cadres, his ignorance of local sensitivities, and his preference for fellow Russians. Certainly he could argue that Kazakhstan was populated more by Slavs and Russians than Kazakhs (in 1979, out of a total population of 14,684,283 inhabitants, 5,991,205 were Russians, 5,289,349 were Kazakhs, and 897,000 were Ukrainian)[19] and therefore needed to reserve a larger place for Russian cadres. In this respect, the decision to name Kolbin was consistent with Gorbachev's speech to the Twenty-seventh Congress, in its emphasis on the often insufficient role of the Slavic minorities in the republics. It was also based on doubts about the competence of the national elites selected based on group affiliations. Hence, it is unlikely that the nomination of a Russian to the leadership of the Kazakh party had been inadvertent. On the contrary, everything indicated that Gorbachev's team wanted greater control over the periphery, particularly the southern and Islamic regions.

By the end of the 1980s, recent developments in this region had Moscow worried. More serious than corruption and criminality, which the center knew were not the exclusive property of the Muslim republics, was the combination of Islam and local sociopolitical traditions. The influence of tradition continued to grow in this whole border region, which adjoined a Muslim area that was being propelled by the Khomeini revolution into a fundamentalist whirlwind. Social relations, authority, and economic ideas tended to deviate from the Soviet model and settle into other molds. Kunaev's behavior during the early 1980s has already been described: when he sat in the Politburo in Moscow, he behaved like a Soviet and expressed himself like a Soviet; in his own republic, however, he reverted to clan head and found other alliances and other modes of authority. The distance from Moscow and the immovable power of Brezhnev—basically indifferent to realities and concerned only with appearances—had encouraged such a revolution. It was all the easier for Kunaev to find this type of authority in his republic since he was powerful in Moscow. And in a Muslim setting, Iran's Islamic revolution reinforced the certainty that this "de-Sovietization" of authority and society was a way to participate in the great movement stirring up the entire Muslim community. Where Moscow saw only reprehensible resurgences of a vanished world, the populations of Central Asia increasingly interpreted this development as a reconciliation with their identity. A profound misunderstanding of this situation already existed when Gorbachev reached power.

All things considered, the appointment of a Russian to this "reserved" post could not have been more insensitive. Recent demographic developments in Kazakhstan neutralized the argument for a strong Russian population. After long being a minority in their own republic, not just in absolute figures but also in relation to the Russians, and after a seemingly inexorable decline, the number of Kazakhs shot up during the 1960s and especially the 1970s, evidence of an unforeseeable demographic dynamism. By the late 1980s, this rise led to a reversal of past trends, since for the first time the Kazakhs clearly outnumbered the Russians (6,531,921 Kazakhs, 6,226,000 Russians).[20] From then on, because of a higher birthrate and a young population, the Kazakh hope of someday becoming the majority group in the republic was realistic. Although Moscow took little notice of this development, the Kazakhs themselves were aware of it and drew from it a legitimate sense of self-assurance.

Because it did not know about these new population trends and their meaning, the central government considered that Kazakhstan was the ideal place to try a change in the leadership policy, modify the distribution of responsibilities hitherto in effect, and, in the confusion thus created, regain control of the worrisome population by a more Russianized leadership.[21] In fact, the central government imagined that in distant Kazakhstan, with its strong contingent of Europeans, it would be easy to implement a new cadre policy and that this precedent then could be extended elsewhere.

This was a disastrous miscalculation that, by ignoring the actual situation in the republic, provoked the Soviet Union's first major insurrection since the riot of Novocherkassk in 1962. In 1962, however, the Soviet system was powerful enough and the news sufficiently surpressed that nothing came of this tragedy and the hundred deaths it cost. In 1986, the central government could not silence the events that shook the USSR, and this first insurrection was to have momentous consequences throughout the country.

The Kazakhs were then to prove that they did not accept the analysis of corruption as decreed by Moscow. They would restore order in their republic, but they refused to admit that the good was in Moscow and evil was on the periphery and that the center therefore could set itself up as role model. Here, as in Tashkent or Tbilisi, the struggle against corruption did little to restore morale. It doubtless involved a continual changing of leaders, which hardly helped the economy. On the contrary, perestroika required a certain stability in the leadership. The prevailing instability undermined all systematic efforts to establish genuine authority. In the growing disorder, once again the gangs and traffickers ruled while cadres, knowing their jobs to be precarious, preferred to submit to the authority of the proliferating "godfathers."

This failure of the effort to introduce higher ethical standards in public life[22] was accompanied by the more serious failure to regularize relations among ethnic groups. Ignoring or rejecting the moralistic directives from the central government, the nations on the periphery resented them as no less than insults that attempted to put them in an inferior position. And hostility toward the Russians—since the whole central government was dominated by them as well—could only increase. At the periphery life was hard, and the necessary but inept reorganization precipitated a chain of suspicions and hatreds whose only possible outcome would be violence.

Part Two

Explosions

3

"Kazakhstan for the Kazakhs!"

POLITICAL demonstrations in the USSR were, quite simply, impermissible. The last mass demonstration by Soviet citizens occurred in Moscow on November 7, 1927. That day, assembled against a government in which Stalin already occupied a central position, nearly all of Lenin's companions, the heroes of the revolution, were dispersed by nightstick-wielding police and then accused of trying to seize power from the Party. The punishment for their infamy was exemplary: in time, they all were killed.

The lesson was not lost. No one demonstrated in the country of workers who, since they held the power, had no one to demonstrate against. Those who risked doing so were obviously enemies of the people.

Decades followed of public calm. In the early 1960s, however, the West heard about uprisings in some Russian cities triggered by a hike in the price of bread. But that received little attention because the silence prevailing over events in the USSR made any information hard to evaluate.[1] The idea thus persisted that Soviet citizens were disciplined, passive, and accustomed to submitting without fuss. All Soviet leaders had long subscribed to this idea. Hence their surprise and confusion when serious rioting broke out in Alma-Ata, the capital of Kazakhstan, on December 17, 1986.

A few months after the glasnost was arduously implemented, following the Chernobyl catastrophe, it seemed that finally there was widespread agreement on the need to make public the events affecting the country. The riots in Alma-Ata would make it possible to measure the progress achieved in this area. Progress was far from outstanding. On December 18, TASS announced that riots had disturbed the capital of Kazakhstan for forty-eight hours but had been suppressed. TASS and then other sources supplied meager information. Young people armed with iron bars and paving stones had gathered on the main square, shouting "Kazakhstan for the Kazakhs and only the Kazakhs!" Were these acts of vandalism, or was this student demonstration that went out of control? Earliest stories were sketchy. Were there several hundred or several thousand demonstrators? The few foreign journalists were forced to dig hard to uncover the facts. Nearly 10,000 demonstrators, many of them college and high school students, chanted nationalist slogans. The objects of their demonstration were public buildings, primarily Party headquarters. Facing them were hastily deployed army troops, reinforced by tanks and other armored vehicles, that brutally suppressed the demonstration. The final toll of injured and killed has never been determined. Official reports mentioned two deaths—a demonstrator and a soldier; the Kazakhs claimed many more. Certainly, many demonstrators were wounded, and many were arrested.

After calm was restored on the evening of December 18, Moscow dispatched to Alma-Ata a member of the Politburo, Solomentsev, who led the investigation but did not make public the facts or his assessment. This silence was prolonged,[2] as was the silence surrounding the estimates of the number of deaths and the number of demonstrators. The riots in Alma-Ata received no more attention from the press in Moscow than they did in the local press or that of Central Asia. It was officially a minor event, but for the first time in a long while in the USSR, a large demonstration took place and was publicly talked about.

In the USSR of 1986, the resort to military and police force and the repression of the young demonstrators contradicted Gorbachev's speech. Letters to the editor in the newspapers expressed little desire for public information, however, a prevailing indifference that was due in part to the great geographical distances and the speedy return of calm. By the time these demonstrations were announced, they already belonged to the past and order had been reestablished. The speed with which they were put down probably suggested their unimportance, but in fact it mainly reflected the lack of preparation and inexperience of those who took to the streets. Demonstrating was too rare an event in the USSR of 1986 for demonstrators to defend themselves and nurture their movement. This probably also explains why there were not many victims.[3]

On the other hand, there was no dearth of explanations about the causes of this crisis. The riots were the doing of alcoholic and drugged "young hooligans" and backed by "nationalist extremists." Furthermore, the official explanation was that everything had been provoked[4] by the supporters of the former and recently deceased first secretary of the Kazakh Communist party, Kunaev. His relatives, collaborators, and clan, who had benefited from the corrupt practices to which Kazakhstan was prey under Kunaev's leadership, could not accept the loss of their privileges. They unleashed volatile young people into the streets to show that, without Kunaev, there would be disorder. As biased as this explanation was, when combined with the idea that it was simply a demonstration by hooligans, it touched on the truth, not because of a so-called conspiracy but because it stressed the political changes then taking place in Kazakhstan. They were indeed the trigger.

No to Russian Administrators!

The immediate source of the Alma-Ata crisis was without any doubt the appointment of the Russian Kolbin to replace Kunaev.[5] The problem lay in his nationality, which was immediately resented in the republic. Moreover, the central government soon realized that even this republic was an unpropitious setting, even momentarily, for its attempted Russianization of the party leadership. On January 10, 1987—that is, three weeks after the riots—ethnic balance was reestablished at the head of the party. The second secretary, who had always been Russian, ceded his post to a Kazakh.[6] All the same, old habits had been too disrupted, and it was the Russian first secretary who sat in the Politburo in Moscow; repression was symbolized by the Russian Kolbin. The population of Kazakhstan did not consider itself free of the central government. Nevertheless, it had to wait two and a half years to see a Kazakh once again at the head of the republic's Communist party. Nursultan Nazarbaev, the man who was named to that post in June 1989, enjoyed a good reputation in both Moscow and Kazakhstan.

The Kazakhs' opposition to Russianization thus seemed to prevail but with a delay of two years, which illustrated Moscow's poor comprehension of local reactions and the republic's problems. When they shouted, "Kazakhstan for the Kazakhs!," the demonstrators meant, "No Russian stranglehold on the republic!" This acute national sensitivity, the first to appear in Gorbachev's USSR, had two very understandable causes: the republic's demographic situation—a source of anxiety for some and hope for others—and the Kazakhs' material situation.

"We Are the Most Numerous"

As often in the USSR, the sense of nationality first asserted itself in demographic statistics. Kazakhstan was long considered an aberration of the nationalities policy, one fated to disappear. According to this policy, every republic in the USSR was characterized by the existence of a national group that was in the majority. Kazakhstan, a land of Russian colonization, did not conform to this policy. The Kazakh people have always been a minority in their own republic, outnumbered by the Russians who lived there. Table 3.1 indicates demographic trends in Kazakhstan, based on five of the more than twenty national groups living there (some of which comprise no more than a few hundred individuals).

The table is significant in many respects.[7] It reflects the long minority position of the Kazakh people. It also reflects the reasons for frustration that, at a time once everything could be expressed publicly in the USSR, acquired an explosive force. The Kazakhs were in the minority because they were physically exterminated in the 1920s, when the Soviet government's wish to stamp out nomadism (can a Soviet be a nomad?) cost this group a million deaths—a quarter of its population.

To the memory of genocide used to enforce a policy of settlement, the

TABLE 3.1
Demographic Trends in Kazakhstan

	1959	1970	1979	1989
Kazakhs	2,795,000	4,234,000	5,289,400	6,531,921
Russians	3,974,200	5,521,900	5,991,205	6,226,400
Germans	648,000	858,000	900,207	956,235
Ukrainians	762,000	933,000	897,964	895,964
Uzbeks	137,000	216,000	263,300	332,000
Total Population	9,924,000	13,008,000	14,684,300	16,463,115

Kazakhs added their grievances about the policy of their colonization. Russians, Ukrainians, and Belorussians were the colonists who had overrun their lands, forcing the indigenous Kazakhs out of the most fertile part of the country to the more arid south. In addition, Stalin used this republic as a detention center for the national groups whom he "punished" for collaboration (Tatars and the people of the Caucasus) or for preventively displaced peoples (Germans from the Volga). In the Kazakhstan that eventually became a national mosaic, the minority Kazakhs (at the start of the 1960s, they represented less than a third of the republic's population) should have integrated with the more numerous Russians, who were the assimilating force for the government. Nevertheless, they held onto their republic; but the presence of the Russians, both in Moscow and in their own republic, left them with few illusions about the national character of the state that bore their name. Kazakhstan long resembled a simple extension of Russia.

For decades, the Kazakhs had nothing to rejoice about—humiliation because of this acknowledged minority status, anxiety about the decreasing population (3,800,000 Kazakhs were counted in the 1897 census, and only 2,900,000 in 1934, while everywhere the rest of the population was increasing), and the painful memory of the victims of settlement. And suddenly in the mid-1980s, coinciding with the Gorbachevian turning point, the Kazakhs noted the end of a seemingly inevitable population decline and even a reversal of the trend. In the 1989 census, the number of Kazakhs surpassed the number of Russians and was increasing rapidly: not only had Kazakhs become the largest national group, but they were on the way to forming nearly half the republic's population.

On this point, Soviet leaders had too long been blind. The central government did not understand the bitterness of people who felt their identity and survival threatened. They did not hear the cries of triumph in the mid-1980s when Kazakh intellectuals observed that the demographic trend had shifted and that their hope of leaving the Russians far behind at the turn of the century was becoming a certainty. All the humiliations vanished in favor of an idea: Kazakhstan was again becoming Kazakh and must return to the Kazakhs.

At this time Kazakhs began thinking of their demographic advance in terms of power—if they were the most numerous, they must hold the largest share of posts of authority and decision-making—but Moscow decided to Russianize Kazakhstan's communist apparatus. Certainly, the demonstrators of December 1986 had not had a clear idea of the demographic breakthrough or taken to the streets because of the census figures. But the observation of this demographic progress—which was decisive for national pride—was a major factor in shaping public opinion and in the nationalistic agitation that then flared up. Because it ignored these developments, the central government had to deal with this first large popular demonstration and opted for a brutal repression that in a few hours tarnished Gorbachev's image in that part of the USSR. He was no longer perceived as a man of renewal but as a traditional Russianizer and the instigator of repression.

Every Man for Himself?

More numerous than before and more numerous than other ethnic groups—particularly the Russians—the Kazakhs also realized that they were among the most economically disadvantaged people in the federation and that, in the USSR's declining years, their impoverishment was rapidly worsening.[8] Despite its enormous potential, Kazakhstan was in 1990 among the three poorest of the fifteen Soviet republics, just ahead of Tadzhikistan and Turkmenistan.

In economic development, Kazakhstan in 1970 was overtaking Moldavia, Armenia, Azerbaijan, Kirghizia, Uzbekistan, Tadzhikistan, and Turkmenistan. Although in 1985 it was still ahead of the four latter states and was the most developed republic in Central Asia, it was only a relative success, for, starting from the bottom, Uzbekistan and Kirghizia were progressing more rapidly than it was. There were certainly abundant reasons for this spectacular lag. Kazakhstan was a major victim of the economic specialization that was too long fashionable in the USSR. As long as the Soviet economy invested in the heavy industry needed for development, Kazakhstan benefited from Moscow's attention and investment. It also benefited when Stalin's successors tried to expand cultivation in order to resolve the eternal problem of Soviet agriculture. When new territories were opened for settlement, Kazakhstan became the "new frontier," and the stampede for Kazakh land brought considerable wealth into the republic. But these policies had run out of steam: attention paid to the land, which had become exhausted through overplanting, had been to the detriment of other regions. Far from Moscow and lacking skilled workers, Kazakhstan did not seem worth industrializing to planners looking for new projects for an exhausted economy.

Abandonment and decline do not explain everything. Impoverishment is a fact common throughout nearly all the USSR. But Gorbachev's economic plans were a particular source of anxiety for the inhabitants of this republic. Until 1985, the Kazakhs could be content with a declining economy, for the traditional system of transferring resources enabled Kazakhstan to benefit from the relative prosperity and efforts of wealthier republics. Certain republics had sufficient production and distributed their income in a balanced way (this was the case in Russia, the Ukraine, Georgia, and so on); others clearly lived below their means, such as Armenia and Azerbaijan; the poorest were assisted. Of all the republics of the USSR, Kazakhstan lived most beyond its means, benefiting from transfers of resources that represented nearly a fifth of its national income. Worse yet, it lived in a growing disparity with its means; the lower its production, the greater the increase in the income consumed. A relatively populous republic, Kazakhstan was extremely costly for the USSR.[9]

As long as no one mentioned this problem, it was of little concern to the Kazakhs. But when Gorbachev thundered against the "parasites" and "profiteers," his announcement that each republic had to become self-sufficient became a source of anxiety for those to whom this fundamental reorientation might be expensive. Even before the texts of his speeches were published, Gorbachev's logic of fiscal autonomy—his speech implied "God helps those who help themselves" regarding the nationalities—caused agitation at the periphery, first of all in Kazakhstan, where its cost promised to be highest.

The fear of a difficult economic future, when Kazakhstan's population was beginning a spectacular rise that would require increased income, explains the mood that prevailed in the republic during 1986. Because the Kazakhs were no longer represented in Moscow and were dominated at home by a Russian party leader, they tended to see Gorbachev's decisions as measures for preserving Russian preeminence in their republic and thwarting their progress.

Internationalism Forever

After the suppression of the Alma-Ata uprising, choosing a policy for Kazakhstan became urgent. The first lesson the government drew from the incident was to avoid fanning feelings of solidarity with the rioters among the population at large—thus, to avoid explaining the reasons for the revolt or attempting to remedy them. The reaction of the Soviet government was far from suited to the character of the event. In Moscow, classic analysis prevailed over deeper reflection. It could not have been otherwise when for months no one—other than Solomentsev—visited the site of the riot, so that the new Soviet leaders had only a vague idea of events in Kazakhstan. It was not until December 1987, a year after the riots, that Prime Minister Ryzhkov visited Alma-Ata. The central government was extremely neglectful of these disinherited borders.

Ryzhkov's visit, which represented a certain change in this respect, was preceded by a debate within the Politburo of the Communist Party of the Soviet Union and a resolution by the Central Committee that finally tried to draft a serious report on the riots.[10] This analysis was still superficial, however. Kunaev and his henchmen were still the guilty parties. Violating the principles of Leninism, they had allowed an unhealthy situation to arise in the republic. They were also responsible for deteriorating relations among the ethnic groups. Up to this point, the main purpose of these attacks seemed to be to bring the Kunaev affair to an end and prepare for his expulsion from the party. Because of these two preliminaries, however, the Central Committee's document deserves some attention. A general lesson emerged: the uprising could only take place because Kazakhstan was losing economic ground, and the solution proposed was obvious—to return to Lenin's idea that one should avoid giving too many opportunities for nationalist feelings to be aired. What united the working people of the republics of the USSR was internationalism; although Kunaev bore some responsibility for the tragic days of December 1986, it was precisely because he had been insufficiently persistent in encouraging internationalism.[11] Furthermore, in passing, the Central Committee's report confirmed that this was not the first attack on internationalism by the young people of Kazakhstan; they had already demonstrated in the city of Tselinograd in 1979. The 1979 incident, which no one could verify at the time, was actually a riot marked by interethnic violence. Students in the city rose up at the announcement of rumors that Tselinograd would be transformed into a German territory to satisfy the demands of Germans whom Stalin[12] had deported in 1941 and had since been deprived of their own territory. In the final analysis, this incidental remark, which accused Kunaev of underestimating the magnitude of the ethnic tensions in Kazakhstan in 1979, illustrates the extent of Moscow's underestimation of them as well. Although the "German plan" was dropped, nothing was done at the time to ease frictions and mistrust among the communities in Kazakhstan.

As in 1979, however, the leaders of the CPSU in 1987 did not notice how likely the coexistence of sorely tried ethnic groups—some because the deportation forced them to live far from their land of origin, others because they felt invaded—might lead to violence. Their response to this situation was familiar and simplistic: internationalism resolves everything, and it was up to local political leaders to foster it.

Internationalism was also the watchword for actions carried out here by Kolbin, who had the difficult task of calming people after bringing about order in the streets. He went about this with doubtful success and a confused idea of the choices to be imposed.

Immediately after the repression, the Kazakh Communist party appointed two authorities to deal with interethnic relations. These appointments confirmed that the party put little credence in the "manipulated hooligans" explanation, but perfectly understood that the crisis had far-reaching origins. On January 9, 1987, the Commission for Interethnic Relations was set up in the Department of Agitation and Propaganda in the Kazakh Communist party for discussing the problems created by the coexistence of various nationalities and suggesting concrete solutions for improving relations between them. The Central Committee of the Kazakh Communist party also appointed a committee on interethnic problems and opened information channels between this authority and the regional party authorities.[13]

At a time when leaders of the USSR continued to believe in the existence of a Soviet people, the creation of these authorities represented unusual progress. But the initiative was local and aroused no particular interest in Moscow. Even then, the gap was widening between on the one hand the central government's perception of events and latent problems and on the other the perception of those people representing it at the periphery. Russian though he was, Kolbin made an effort for several months to work at putting together a profile of the national problem in Kazakhstan.

The first task was to have the commission of the Central Committee assigned to establish a profile of the December demonstrators and to propose a sociological analysis of the Kazakh young people. This investigatory work greatly improved the insurgents' image: the term *hooligans* was replaced by descriptions of real students, as in 1979, who were no longer described as alcoholics or drug addicts. On the other hand, the commission emphasized that these young people's faults partly explained their spirit of rebellion: the poor education they had received reflected a mistaken policy of national advancement.

The Soviet government had indeed encouraged the creation of national elites within the republics in a way more or less comparable to the American policy of affirmative action for the advancement of African Americans. By relaxing entrance requirements met by Russians admitted to institutions of higher learning, the government had tried to form large masses of non-Russian students by lowering academic standards. This policy had unfortunate side-effects. The universities were saturated with too many Kazakhs: from 70 to 90 percent of the students at the University of Alma-Ata were Kazakhs, who represented only 40 percent of the total population. This mass of underqualified students brought down the general level of studies and created a homogeneous national milieu, far different from the desired heterogeneity of the Soviet people. Selected according to undemanding qualitative standards, these students also benefited from the advantages of "national preference." Onto the federal program for the "advancement of nationalities" had been grafted a local program for the "advancement of compatriots," whatever their educational level.[14] And this level was appalling.

These students, often inept at their studies, also were inept at jobs that required genuine competence. The gap between a mass of graduates and the real needs of the republic produced frustrations, of which the rebellions of 1986 may have been a symptom.

The linguistic shortcomings of these students provided more evidence of an education ill-adapted to the needs of the USSR. The Central Committee noted that one of the major educational failures in Kazakhstan was its negligent propagation of the Russian language, which was indispensable for interethnic communication. Moreover, it was no more successful in teaching Kazakh to those who claimed they had a good knowledge of it. No doubt the investigators took some satisfaction in claiming that the young Kazakhs who demonstrated at Tselinograd or Alma-Ata to defend their national rights were not capable of expressing themselves correctly in their own language. Beyond its spitefulness, however, this observation alluded to a subject of great Kazakh discontent—that the Soviet educational system viewed their language with disfavor and that it condemned the Kazakhs to constantly resort to speaking Russian.

The commissions set up by Kolbin were not without effect. By opening the Pandora's box of linguistic conflicts, they allowed the Kazakh population to understand itself better, to obtain a hearing, and to hope that a policy that took account of its aspirations would allow for greater protection of its national interests. The Kazakh press could finally make this paradoxical situation public: in the republic, the Kazakh language theoretically had the same rights as the Russian language, and the logical corollary of this status was formation of national elites; in practice, however, the language of public life had become Russian. Examples abounded: no part of the administration understood applications drawn up in Kazakh or had any forms printed in that language; the bureaus—economic, social, and so forth—where Kazakhs worked had only a few typewriters with characters used in the Kazakh language; the typists trained to use Kazakh had to yield to Russian or Russified personnel; audiovisual materials in the schools were in Russian; and so on. The commission noted that, before any policy for the advancement of national elites, the advancement of the national language in public life was obviously needed.

For a Russian leader—in this case, Kolbin—these findings made it all the harder to determine a clear line of demarcation. Although he encouraged the investigative efforts of the commissions created after the riots, he later tried to maintain an even balance between national aspirations and central interests. On the one hand, he became an advocate of Kazakh progress and admitted that Russian cadres—and, more generally, all the Russians—had to stop speaking only Russian, which created the impression they were living in a conquered country. For this was how things seemed: although in 1989, 62 percent of the Kazakhs said they were bilingual (which is probably correct, for how else would they have survived in a republic dominated by the Russians?), only 0.9 percent of the Russians knew Kazakh.[15] Kolbin denounced the contempt of Russians for the country in which they lived and set an example by promising to change his own behavior; he committed himself, before audiences that were sometimes cordial and sometimes sarcastic, to learn the republic's language in one year. On the other hand, he was irritated by some of the demands of the Kazakhs in this matter. When they asked that kindergartens and day schools be opened to educate their children in the national language, he impatiently retorted that progress was not determined in kindergartens and that if Kazakhs desired effective participation in public life, they would learn Russian instead of having only a rough knowledge of it.

In April 1987, a meeting on interethnic relations was held in Alma-Ata to discuss these problems. It was evidence of the good-faith desire of the government to balance two contrary positions—that of the Kazakhs, who demanded the advancement of their language, and that of the Russians, who judged that they were discriminated against in the universities and in public employment where the Kazakhs continued to enter under relaxed admission standards. By making pledges to both parties, Kolbin left them both dissatisfied and sowed the seeds for new conflicts.[16]

Was this done to subject the region to the policies of the center? Was it done simply because the influence of the two opposed communities was balanced? Or faced with this dilemma, did Kolbin's conclusion amount to more a pious wish than a realistic policy? Everything is resolved in internationalism, he unendingly repeated.

Internationalizing mentalities is no easy task. To achieve it, Kolbin outlined an immediate step—the internationalism of jobs and the workplace.[17]

Russians and Kazakhs have long led separate lives. The traditionally stockbreeding Kazakhs do not participate in the industrial world. The community feels that skilled and unskilled workers lack prestige. The Kazakhs who live in urban areas prefer service jobs or administrative posts in factories. They thus avoid mixing too closely with Russian workers and can use their own language among themselves, even in their professional lives. Once again, the linguistic problem and a particular perception of the job hierarchy helped keep the communities separate, even in the cities, which traditionally have encouraged ethnic intermingling.

Economic development requires that the republic increase its efforts toward industrialization, which could provide jobs for the growing local labor force and eventually create in the working world a community transcending national differences. After the Alma-Ata riots, the local government under Kolbin's leadership urged young Kazakhs to take jobs in industry. This proved to be a waste of time: young people complained that to persuade them to take jobs in industry, the government barred them from entering the universities and intended to replace them with representatives of other communities. Diplomas in hand, they wanted only posts of authority; when they did not graduate, they simply refused to take jobs that would involve mixing with Russians.

The catastrophic state of the Kazakh economy—where housing was even scarcer than it was in the rest of the USSR—and the vicissitudes of supplying foodstuffs were also the subject of a crossfire of accusations. Kolbin criticized the Kazakhs for refusing to contribute to the development effort; the Kazakhs claimed that, whatever happened, the Russians' privileges doomed any development policy to failure.[18]

Although the deeper causes of the 1986 uprising were not well understood in Moscow, they were in Kazakhstan. The rapid development of national feelings in the republic and the sudden boost provided by the Kazakh victory in the demographic battle (they were already thinking of the next millennium and figured that they would then number some 12 million) combined with the fear of reforms and the growing hardships of daily life. There was nothing new about this picture; it more or less held true for many of the peripheral nation-republics of the USSR. But new in Kazakhstan was the sudden and violent eruption of ethnic factors in political life. When Moscow imposed purges at the periphery to reform corrupt practices and sent in leadership cadres without real safeguards, local discontent festered. Until 1986, however, it had never been expressed in the form of an outright conflict between Russians and non-Russians. People blamed the central government, its interventionism, and the unequal relations between the center and the periphery—a general theme that certainly disguised profound interethnic resentments. But new at Alma-Ata in 1986 were the undisguised expression of anti-Russian feeling and the assertion that Kazakhstan must belong to the ethnic group representing the majority of the population.

The optimistic discourse concerning the Soviet people—internationalism and its progress—had been broken up in 1986 by the brutal slogan "Kazakhstan for the Kazakhs!" Through an irony of Soviet history, the 1986 rupture occurred in the least likely republic—which no doubt explains the inept decision making of the central government. The numerical weakness of the Kazakh national group, the large size of the Russian community, and the many nationalities living in the area all suggested rapid acculturation, if not assimilation. Nowhere else in the USSR was there a greater likelihood of seeing various ethnic groups melt into a Soviet people.

Two elements could allow this development to be foreseen. First, the central place of the Russian language in the life of the republic was criticized by the Kazakhs.[19] Successive censuses confirmed that Kazakhstan was unique in this respect. In Central Asia, only 25 to 30 percent of the people in the nations with republics of their own—Uzbeks, Tadzhiks, Turkmenians, Kirghiz—knew Russian. In the Caucasus, even among the better-educated Georgians or Armenians, only 30 to 40 percent had mastered the language. The Kazakhs, on the other hand, were among the frontrunners, with the Latvians and ahead of the other Slavic peoples, Ukrainians, and Belorussians, in this race toward linguistic acculturation. Certainly, this support for the Russian language did not result from any desire for an alliance with the leaders of the USSR but rather from necessity. It was required for life in a republic where the numerical strength of the Russians had imposed the near-exclusive use of their language. The Kazakhs did what was necessary for survival. Nevertheless, bilingualism, which was taken to have failed nearly everywhere else and succeeded in Kazakhstan, would suggest that here, at least, the bringing together of ethnic groups was working.

Second, Moscow's illusions were probably fueled by past experience. The republic departed from statistical norms, since the Kazakhs were then in a minority, so Moscow thought that Kazakhstan illustrated the success of the national project. Despite their small population, the Kazakhs preserved an unusual status of shared laws. Certainly, the temptation had sometimes arisen to speed up acculturation and integrate Kazakhstan with Russia, a harbinger of the future unification of the Soviet territory. This was particularly true under Khrushchev. When the latter launched Soviets on an assault on the republic's unsettled territories, he was thinking of incorporating this part of Kazakhstan into Russia and thus of annexing five of the republic's richest regions, representing a third of its population and two-thirds of its grain production. As a first step toward this incorporation, the capital of the unsettled territories was created in 1956 and its town name renamed and Russianized: the old Kazakh city of Akmolinsk became Tselinograd. This change took into account the population of the territory, which was still mostly Russian. Khrushchev's Kazakhstan policy thus took no account of national sensitivities. Political staffs were massively Russianized, and the possibility loomed that all Kazakhstan would be reduced to the rank of an autonomous republic or territory attached to Russia: after all, the republic of Karelia had experienced such a fate.

The threat that then hung over Kazakhstan's future was far more serious than the "parachuting" of a Russian bureaucrat to the post of first secretary of the republic's Communist party three decades later.

No doubt, like Khrushchev, Gorbachev thought that Kazakhstan was the most propitious place for promoting the internationalist ideal of the USSR. But it is testimony to the depth of change in public opinion that although Khrushchev's plans had not stirred up Kazakhstan, a riot made Gorbachev back down definitively. Khrushchev's project collapsed not from the blows of antagonistic Kazakhs, but because the poorly conceived and ineptly carried out conquest of the virgin lands proved costly and futile. Khrushchev's successors gave up the whole plan. Then governmental stagnation cast dreams of territorial and political transformations into oblivion. Reassured about their future, Kazakhs gradually regained the leadership posts lost earlier to Russians. Of the threat that hung over their future as a state, there remained only the name Tselinograd.

It may have been this memory of the calm with which Kazakhstan had greeted Khrushchev's resettlement policies that helped convince Gorbachev that the only problem to resolve was the elimination of corruption and favoritism. But the systematic advancement of the national elites—a Soviet variant of the American policy of affirmative action—which had gone on since Khrushchev's downfall, in fact encouraged the very phenomena that Gorbachev wished to eliminate. It also weakened the authority of the central government and of the Russians in the republic. Like Khrushchev before him, what Gorbachev tried to test in Kazakhstan in 1985 was the cessation of affirmative action and the reality of a Soviet people who were no longer separated by ethnic barriers. The violence unleashed in response to this test—the sudden mobilization of a whole society—contradicted all earlier certainties. Far from uniting people, decades of common life, bilingualism, and the "generosity" that formed the national elites and granted them a privileged political status all helped created antagonism. This was the cruel and unexpected lesson of the first great popular uprising. This lesson was understood by all within the borders: Alma-Ata revealed that, in Gorbachev's USSR, it had become possible to rise up against Moscow. This lesson was not learned at the center, where attention quickly turned away from the peripheral regions and everything was explained by antisocial individual behavior and the consequences of stagnation.

In 1913, a newspaper published in Kabul with the headline "Asia for the Asiatics" helped hasten national awareness in Central Asia. In St. Petersburg, no one noticed it. In 1986, the slogan "Kazakhstan for the

Kazakhs!" similarly resounded and with the same effects—a warning to the center and an incitement to the periphery. Here, history tends to repeat itself. In Moscow, no one worried about the possible contagiousness of this slogan, while it endlessly resounded in neighboring republics. In a few months, it could be observed that, as in 1913, by ignoring the grumbling that shook its borders, the central government was compromising the future of the empire.

4

The Lebanonization of the Caucasus

BEFORE February 1988, few outside the USSR and Armenian or Azerbaijani circles had ever heard of Nagorno-Karabakh. Nevertheless, it was the fate of this autonomous region, an enclave within and united with the republic of Azerbaijan, to show the world the magnitude of the interethnic conflicts in the USSR, plunge the whole Transcaucasus into civil war, and prove that Soviet federalism had run out of steam.

The crisis of the Transcaucasus has many dimensions, including the problem with Nagorno-Karabakh, the environment, the relations between Armenians and Azeris, the growing hostility of the two nations toward Moscow itself, and (since the Azeris are ethnically Turkish) relations between the USSR and neighboring Turkey. Each alone would be enough to fuel the conflicts of a violence hitherto unknown in the Soviet Union.

"One People, One Republic"

Armenian separatism in Nagorno-Karabakh was the trigger that catapulted the whole Transcaucasus from seeming tranquility into horror. Covered over for decades in the USSR, this ethnic problem had in fact existed since the beginning of Soviet power.

A mountainous region of the Caucasus that in its current administrative form covers about 2,740 square miles, Karabakh had been the object of permanent conflict between Armenians and Azeris, who demanded possession.[1] In December 1920, when Armenia lost its independence, it was tacitly admitted that its borders were to include Karabakh, whose population was 95 percent Armenian. The desire to satisfy Kemalist Turkey, however, led the Soviet government to renege on this commitment and in 1923 to cede Karabakh to Azerbaijan. For Armenians of both Armenia and Karabakh, this situation was hard to accept, especially since within Azerbaijan, Karabakh had only the status of an autonomous region, dependent on the political choices of the surrounding republic. The tormented history of the USSR from 1923 to the death of Stalin prevented Armenians from dwelling on their misfortune. In addition, the horrifying memory of the genocide of 1915 made Soviet protection important. Armenians thought that their country was safe from any repetition of such an abomination and thus had to accept the drawbacks of this protection.

The return of national aspirations was connected with political developments in the Soviet system. The end of Stalinism put an end to fear to some degree and revived thinking about autonomy. The nations of the USSR once again turned their attention to their particular destinies and sorted out the gains and losses of their association with the USSR. Armenians gradually recovered their memory. The memory of the genocide led them to wish to celebrate the memory of the victims—but the claim of genocide was disputed by the Soviet government. At the end of the 1950s, when Khrushchev's USSR wanted a reconciliation with the successors of Mustafa Kemal, dwelling on the events of 1915 amounted to interference with its international designs. The silence imposed on this memory contributed little to the awakening of national feelings in Armenia.

In Karabakh, this rediscovered memory fueled a never-suppressed separatism. In 1965, the fiftieth anniversary of the genocide intensified the demand for "one people, one republic." For nearly a quarter of a century, the Karabakh population pleaded their case with Moscow for making them part of Armenia. Among the arguments put forward was cultural discrimination and Azerbaijan's economic desertion of this region, which in 1979 had 162,000 inhabitants and ten years later 185,000, of whom 145,000 were Armenians and about 40,000 Azeris.[2] It was easy for Karabakh's Armenians to stress that their region, which they had pretty much to themselves in 1920, later took in a considerable percentage of Azeris—nearly a quarter of the population. No doubt correctly, they saw in this development a systematic desire to invade Karabakh, weaken the region's ties to Armenia, and affirm the legitimacy of its annexation to Azerbaijan. This gradual change in the balance of the population induced Armenians to ask more insistently than ever for Karabakh's annexation to Armenia.

Moscow remained deaf to the plea for reconsidering the borders of 1923. This ignoring of a real problem helped account for the rapid escalation of Armenian fury.

At the end of the 1980s, as in 1956, political developments in the USSR brought the debate to the public area and in violent terms. Did not perestroika imply that everything in the USSR was up for reexamination, rethinking, and, if necessary, reforming? That, at least, was the belief of Karabakh's Armenian population, who presented Gorbachev with a petition signed by nearly 80,000 people (the entire adult population of the region), a genuine popular referendum asking that the borders be redrawn. A year of silence was followed by a disdainful, negative reply that came from the administration of the Central Committee of the CPSU, although the appeal had been addressed to Gorbachev himself. Once again, the secretary-general of the CPSU showed his incomprehension of the feelings of people at the periphery.

In Karabakh, the appeal's rejection sparked an instantaneous mobilization: 1988 was not 1965 or 1975. The USSR had changed, and the demonstrations that were multiplying at the periphery had taught the population that, to make its cause prevail, it had to express its will en masse and refuse to accept that any reform depended simply on the good will of those who governed them. On February 11, 1988, demonstrations took place in Stepanakert, the capital of Karabakh, confirming that the discontented masses were becoming a participatory civil society.

The street parades of Stepanakert were echoed in Yerevan, for in Armenia too the people were losing patience with being ignored. In addition, solidarity with Karabakh was here coupled with another cause for agitation—anxiety about the environmental future of the republic.

Until the explosion at Chernobyl in the spring of 1986, no one in the USSR admitted that serious environmental problems could exist there. Stalin's old refrain "Nowhere do people breathe more freely than here" had changed from its initial political meaning to an ideal vision of the environment. Pollution and the dangers of rampant industrialization concerned only the capitalist world; the USSR was spared them. Certainly, at the beginning of the decade, Soviet scientists and various other groups had begun sounding the alarm: their country was the site of an unprecedented ecological disaster. Before he was exiled, Solzhenitsyn had addressed the Soviet leaders with a moving appeal on this theme, warning against "our collapse and that of Western civilization in the chaos and stink of the sullied earth."[3] After Chernobyl, glasnost opened a flood of tragic information, but with no practical results. Starting in 1988, however, the Soviet government, horrified at discovering the extent of environmental destruction and its consequences for public health, allowed information to be published about the dangerous cities.[4] Yerevan, the capital of Armenia, was among the most polluted cities of the USSR,[5] at least as regards carbon monoxide. The pathological consequences—miscarriages, deformed newborns, respiratory illnesses, cancer, a higher mortality rate—were so numerous that they were the subject of official investigations, including an examination of classified documents.

Armenian intellectuals mobilized and called on Gorbachev to cease blanketing Armenia with polluting factories. This proved to be a waste of time. In 1988, work began on a new chemical complex. Demonstrations multiplied at Yerevan, where a third of the republic's population felt threatened with "environmental genocide." In February 1988, from Stepanakert to Yerevan, tens of thousands of Armenians shouted in unison, "Give us back our unity! Let Armenia live!" It was not surprising that the combined national and ecological demands lent an extraordinary strength to the movement. At issue was the survival of a people threatened with physical destruction and with the deluge of Azeris. Armenia was felt to be too small a region with too small a population to cope with this twofold peril.

In these early days of the "Karabakh revolution," the adversary was an Azerbaijan that refused to negotiate about the future of the region. In the years of Soviet change—the "Gorbachev years"—Azerbaijan had hardened its positions. Perestroika had few echoes, but Azeri national feelings about its past and culture demonstrated a desire to benefit from the intervening developments in the USSR to strengthen the identity of the Azeri nation and its life in the republic. At a time when Armenian nationalism was burgeoning, Azeri national feeling, which always had kept its hostility toward neighboring Armenia, could clearly only fuel the desire to have done with this hated neighbor.

The effects of perestroika in Moscow added to this open hostility. In the fall of 1987, Geidar Aliev, who until 1981 was the first secretary of the Azeri Communist party before becoming first vice president of the Soviet government, was dismissed from this prestigious post and stripped of his seat in the Politburo. The Azeris felt that they had lost their representative in Moscow just when Gorbachev was surrounding himself with Armenian advisers. They dreaded seeing him favor the Armenian claims. Their demographic advances also helped precipitate a more assertive national climate than before. This republic had a population of some 7 million in 1989[6]—up from 6 million in 1979—and the population growth was due solely to the Azeris. In ten years, the Russian population fell from 475,000 to 392,000. Beyond the increase in numbers in Karabakh, the Azeris almost totally populated Nakhichevan (281,000 Azeris and less than 2,000 Armenians out of a total of 294,000 inhabitants in 1989), the area that the territorial carving up of 1923 removed from Armenia and placed in Azerbaijan. This policy of systematically populating areas showed that Azerbaijan did not intend to lose an inch of territory or to consider for a moment any idea of redrawing borders or even an exchange of populations between Nakhichevan and Karabakh. In Baku, the demonstrations in Stepanakert and Yerevan were followed with passionate interest, but with the firm intention of yielding nothing. As muted as it still seemed, conflict was ready to erupt between two nations that were soon to tear each other to pieces. The ingredients had not changed since 1923, but since then, passions had intensified on both sides.

At the beginning of 1988, the center of the system was not the target of any hostility. The Armenians, still haunted by the dreadful memory of 1915, did not think of relinquishing Soviet protection. They waited for arbitration from Moscow, even if the silence that prevailed there in the face of repeated demands for a revision of the borders seemed unencouraging. Azerbaijan similarly held that arbitration from the center, previously so favorable to that republic, was a decisive element in the quarrel over Karabakh. In addition, leaders of the republic were aware that an external element weighed in their favor—Turkey. This country had once intervened so that the Azeri territory was given areas claimed by Armenia. The territorial demands made at the end of the 1980s did not go unnoticed in Turkey and were discussed in the press leading to the resurfacing of the theme that Ankara should take responsibility for protecting the integrity of "Turkish space."

These were the elements of the coming tragedy. Nevertheless, beyond the malaise and the intransigence, nothing suggested that a tragedy was to be played out.

A Programmed Pogrom?[7]

On February 18, 1988, sizable demonstrations occurred simultaneously in Yerevan and Stepanakert. The demonstrators had found the slogan that united them: "One people, one republic!" They were more numerous in Stepanakert as the general strike freed schoolchildren and workers to come out into the streets. But as numerous as they were, there should be no mistake about what motivated them: they had no grievance against either the USSR or the system; they simply demanded that the principle of self-determination, so beloved by Lenin, be applied to them. Their appeal was addressed to the Soviet system as it existed—that this principle written into the Constitution of the USSR be implemented here.

In Karabakh, the popular demands gave birth to a genuine political organization such as the USSR was not accustomed to seeing. On February 20, 1988, after Moscow refused to acknowledge the petition-referendum presented to it, the soviet of the autonomous region voted by an overwhelming majority (110 votes for, those of all the Armenians, the thirty Azeri deputies having boycotted the balloting) for the annexation of this region to Armenia. This was a major first in the Soviet Union: deputies elected by undemocratic balloting suddenly discovered themselves representative of the popular will and no longer the mouthpieces of the central government. That day, modern democratic politics was born in the USSR as suffrage suddenly acquired meaning.

Just as politically important in Armenia was the appearance of the Karabakh Committee, a popular front that preceded similar fronts that were to emerge elsewhere, and in a few months it became the real power in the Armenian nation. The vote of the Karabakh regional soviet, the formation of the Karabakh Committee: these elements were to bring the Armenian people onto the streets to try to impose from below their demands on the central government. In one week, in Yerevan, 6,000 demonstrators demanded recognition for a reunified, purified Armenia that would be restored to prosperity. The demonstrators called on Gorbachev for assistance and on their own parliament to ratify the vote of the Karabakh soviet and thus to decide to place the rest of the Armenian country under its authority. Would civil society win its fight?

On February 26, 1988, at the end of this extraordinary week, Gorbachev reacted. He did so with a plea for calm addressed to the two parties, Armenians and Azeris.[8] He said the times first demanded perestroika and the resolution of the most urgent problems, which were economic and social. He said that the Party understood the difficulties of relations between the nations and was prepared to discuss them. On that day in Moscow, Gorbachev presented two Armenian writers who had come from Yerevan to plead Karabakh's case with an ambiguous message. He claimed to be seeking an "equitable solution" and complained that through its demonstrations the Armenian nationalist movement had dealt a fatal blow—"a stab in the back"—to perestroika.[9] As the author of an excellent history of Soviet Armenia has noted, this speech may be related to General de Gaulle's ambiguous *Je vous ai compris* ("I have understood you") on his coming to power during France's Algerian crisis in 1958.[10] No one really knew what had been understood. When in doubt each person may conclude what he or she likes, and Gorbachev, like de Gaulle, hoped to gain some time and calm inflamed passions.

Far from appeasing them, however, this speech may have hastened the catastrophe—that is, the pogrom. In Yerevan, the promise to create a commission to study the problem, on condition that calm returned, certainly helped the demonstration's organizers to convince their troops to accept the idea of a truce—a month for life to return to normal, a normalcy that was imposed on them despite the reluctance of extremist elements. In Baku, however, things were not the same, perhaps because people feared that under pressure from his Armenian entourage, Gorbachev would accept the opposing demands. And then came the horror of Sumgait.

The week of joyous but nonviolent demonstrations was followed by two terrible days that rekindled the Armenians' memory of 1915. The sequence of events is clear, even if many of the elements remain obscure. It all began began with a "raid" of young Azeris—"angry young people"—in Stepanakert. The clashes, which were inevitable in the overheated atmosphere of Karabakh, left two Azeris dead and dozens wounded. Then came the escalation. The Azeris decided to take revenge for these deaths. The result was the pogrom of Sumgait, an industrial suburb of Baku in which Armenians and Azeris lived together: two terrible days and two horrendous nights of killing, rape, looting, and arson of everything Armenian. The official toll was thirty-one dead and 300 wounded. The reality was surely worse. The events that followed were classic: a curfew, MVD [interior ministry] troops hastily dispatched to the area (but too late), a chain reaction of clashes in various parts of Karabakh and even Baku. Armenians fled Azerbaijan and the memories of horror; Azeris fled Armenia for fear of reprisals. On both sides, the exodus swelled over the months and with repeated violence.

Beyond the violence, the pogrom concealed mysteries. The wildest rumors circulated, adding to the Armenians' passions and despair. Behind these rumors, some strange facts seemed hard to ignore, however. In 1988, while the Soviet Union was still living under the "dry law" that made the purchase of liquor chancy and difficult, vodka suddenly reappeared in the region. Someone had revoked the laws then in force and given permission for an already enraged population to add to political passions those created by immoderate consumption of alcohol. Someone had arranged for bands of Azeris from areas outside Sumgait to be brought in by special buses to the site of the pogrom. Who were these violent young people who seemed to have followed a preconceived plan? No one knew, and the justice system was of little help in sifting the truth from the most fantastic conjectures.

The silence of the central press and the official half-truths—glasnost was not the order of the day—still makes any judgment about the whole impossible. Nevertheless, it is not unthinkable there was deliberate provocation or at least the tacit complicity of the authorities. But which authorities were involved? For the Armenians, the Azeri government conducted or tolerated the frightful raid. For the Azeris, the KGB had the goal of victimizing the Armenians and thus gradually imposing the territorial solution they wanted. Finally, for some of Gorbachev's ardent admirers, since a crime is always the doing those who benefit from it, the culprit was the conservatives who were determined to get rid of Gorbachev.

It is easy to note the effects of the tragedy—a radical change in the problem of the Transcaucasus. Hitherto latent—Armenians and Azeris coexisted in the two republics—the mutual hatred became so strong and overt that it necessitated the absolute triumph of one and total defeat of the other. Coexistence, certainly difficult, had been maintained but was no longer possible. The Karabakh affair could not be left in its existing state.

In both Yerevan and Baku, it was expected that this time Gorbachev would make an unambiguous pronouncement on the matter—which was exactly what he did not intend and no doubt could not do. Taking drastic action would totally alienate one republic and the republics had sources of support that Gorbachev could not overlook. The Armenian diaspora was not forgotten in Moscow, and outside the Soviet Union it mobilized innumerable loyalists. Defying it would amount to defying international public opinion. The situation was different in Azerbaijan but no simpler. Here there was no diaspora, but the Muslim republics in the USSR were waiting for something to rebel against, particularly any disenfranchising of Azerbaijan. Since December 1986, Gorbachev knew that the Muslim periphery could quickly ignite. Beyond Soviet borders, Turkey also looked stormy. On March 12 the daily paper *Miliyet* opportunely recalled that Ankara was accountable for the fate of the Turkish territories, even those located outside its jurisdiction—a subtle threat that Gorbachev could hardly ignore.

When Gaining Time Means Losing It

To gain time, a plan for development[11] in Karabakh was hastily adopted in Moscow. It provoked outrage among Azeris, who said that Karabakh was already better provided for than the rest of the republic. Although the Azeris were displeased, Armenians had every reason to be equally dismayed. They had hoped that the pogrom might force Moscow to listen to their requests. This hope was short-lived. On March 23, 1988, the Supreme Soviet adopted a resolution condemning the Armenian position. The difficult territorial problem of the nations was not to be resolved by pressuring Moscow. The resolution singled out for censure the Karabakh Committee, which was organizing a demonstration for March 26—the grace period negotiated with Gorbachev in February having expired. The Soviet press continued the attack on the committee,[12] and the resolution soon was made public, pronounced by the authorities in both Armenia and Karabakh but clearly issued by the Supreme Soviet of Azerbaijan. Few suspected, however, that the order had come from Moscow. Police and MVD forces endlessly patrolled in Yerevan in helicopters, and four leaders of the Armenian movement were arrested.

The Karabakh Committee was called a dangerous organization that spread disorder and called for uprising, when its leaders, sensing Moscow's hardening, actually sought to prevent street movements and were organizing simple passive resistance, as on "dead city" day in Yerevan, May 25.

Moscow's reaction was hard on Armenians and indicated a certain alarm. The mounting agitation, organizing ability of the Karabakh Committee, and emerging civil society that demanded a hearing for its views were new variables in the relations between the government and the governed in the Soviet Union. Clearly, the government still thought it could check this development and regain control of all decisionmaking. Gorbachev's promise in mid-March to appoint a special commission to investigate the situation simply reinforced the traditional, hierarchical approach to all the problems of the USSR, whatever their urgency and seriousness. For good measure, to show that customary problem-solving methods were not being revised, the two first secretaries of the Communist parties of Armenia (Demirshian) and Azerbaijan (Baghirov) were dismissed. These expulsions occurred with the consent of Gorbachev's closest collaborators, who attended the plenary session of the central committees of the republics convened to review the events of February—Ligachev, a reputed conservative, in Baku, and Yakovlev, an avowed liberal, in Yerevan. Everything was orchestrated to show that at the highest levels of the system, people of all political leanings worked together, and unanimously agreed to settle the problem of the Transcaucasus in a ruthlessly hierarchical way.

The authority shown by Gorbachev was not enough to prevent the unfolding of events. Clashes followed each other for any pretext, and the sentence pronounced in May by the court of Sumgait against one of the instigators of the pogrom—fifteen years' imprisonment—renewed the cycle of violence. For Armenians, this sentence mocked the memory of the dead; for Azeris, the verdict was punitive. As mobilization took place on both sides, Moscow noted that the situation in the streets was becoming uncontrollable.[13] It was even more intractable politically, since on June 15 the Supreme Soviet of Armenia unanimously voted to annex Karabakh, and forty-eight hours later the Supreme Soviet of Azerbaijan voted no less unanimously in the opposite sense. It was a standoff. But more seriously it signaled the beginning of the rebellion of local governments against central government. Restrictions on local parties had not worked, evidently, since Moscow was not seen to be the place where national affairs were decided.

Once again, Gorbachev intervened and clearly opposed any change in frontiers. Armenians still hoped that the party's Twenty-ninth Congress would resolve these problems, but Gorbachev had set the tone. It was expected that the conference would declare that the rights of minorities must be better guaranteed—a vague statement that avoided the issue of

Karabakh. The response of the parties concerned, however, was that on July 12, 1988, the deputies from Nagorno-Karabakh voted to join Armenia. Gorbachev's adoption of these positions and the conference's silence made the vote a fait accompli and declaration of intent. This constituted a revolution in Soviet politics. Once again, the Armenians of Karabakh showed that the right to self-determination must be exercised by the parties involved—and not by Moscow.

At the center, few underestimated the gravity of the situation. Armenians were creating a dangerous precedent: Would other republics or national groups seize the opportunity for self-determination? What would remain of the federation if no one acknowledged the authority of the central government? On July 18, the Supreme Soviet of the USSR met to debate these questions and concluded that Armenia would never acquire Nagorno-Karabakh. Order had be reestablished, and the Supreme Soviet announced that no measure would be spared in imposing it.[14] The Armenians never even had their own proposal to the Twenty-ninth Party Congress considered: they asked that Karabakh be taken away from Azerbaijan and temporarily (to allow time to find a compromise) placed within the framework of the Russian republic.

The year 1988 ended in a succession of violent events, but also in the earthquake of December 7. Estimates of the dead varied from 30,000 to more than 100,000. The number of people left homeless was close to a half million. It was yet another trauma for a nation already severely injured, and the earthquake revealed once again the incompetence and corruption that raged throughout the Soviet Union. Construction standards had been ignored through technical incompetence and skimming of personal profit from public work projects. No doubt many housing scandals all over the world indicate the same human weaknesses with criminal consequences. In Armenia, however, corruption had a high cost: it cost the lives of thousands of innocent people in a catastrophe that convinced Armenians that the Soviet government was responsible for creating an environment inimical to human life.

They had an additional reason for indignation: in the hours following the quake, the central government had the leadership of the Karabakh Committee and numerous militants arrested, probably to take advantage of the population's incapacity during the crisis. A new change of political personnel finalized Armenia's normalization, and for Gorbachev the "Transcaucasian case" was settled. After the Sumgait pogrom, it was clear that one of the two adversaries would have to be disciplined. On December 10, 1988, the combined action of the central government and nature seemed to provide an answer.

Down with the Russians!

As effective as they can be in the short term, solutions by force rarely settle national conflicts. In the Transcaucasus, political or military misrepresentations and violence produced results that Moscow had certainly not expected.

The first change was that expectations centered on Moscow turned into local hostility toward Russia and the Russians. To escape the Turks, Armenians were said to be inclined to side with Russia, and Azerbaijan's xenophobia toward the Armenians led it to be rather lenient toward Moscow and the Russians. Living and working side by side with Russians in industrial centers like Baku contributed to this relative neutrality.

In a few months, however, mutual hostility combined with anti-Russian frenzy to crystallize into hatred. Armenians found everything about Gorbachev's behavior unforgivable. Far from playing its traditional role as protector of Armenians against the Turks, the Soviet Union seemed to be simply delivering them to the Azeris, beginning with those from Karabakh. Russia and the USSR seemed to be the same, for the government was more Russian than it had been for a long time. If Armenians were no longer to be protected from the Turks, then there was no reason to accepts the drawbacks of that protection—subjection to the Soviet Union. Armenia, which up to 1988 was the most docile and accepting of Soviet republics, suddenly succumbed to the charms of emancipation.

On May 28, 1989, demonstrators paraded in Yerevan to commemorate the seventieth anniversary of independence, waving the red, blue, and orange flag of Free Armenia. The three colors inspired them to demand their own destiny. They were not yet dreaming of absolute independence outside the federation but wished to put an end to Russian domination, enter into a truly egalitarian system, and struggle to impose it in alliance with the other national movements that were asserting themselves. Armenia no longer wanted a protector but sought partners. Soviet federalism had been badly hurt.

Azeris were no less hard on Moscow, although they had reasons to be content. No one in Moscow challenged their right to hold Karabakh. But they were enraged at measures they felt were concessions to Armenians, although they themselves wanted an "all or nothing" policy. They felt that Karabakh came completely under the sovereignty of Azerbaijan. Under these conditions, they could not accept Moscow's insistence on the cultural rights of the Armenians while Azeri policy was for a massive influx of Azeris that would overwhelm them in a Turkish-speaking majority. Neither could they accept Moscow-appointed commissions that encouraged Armenians to file complaints and defend their rights.

Azeris found Moscow's meddling in favor of the Armenian minority intolerable because it represented an attack on the republic's sovereignty, which was written into the constitution. The Azeris imputed this meddling to the cultural solidarity that united Russians and Armenians around Christianity. Muslims versus infidels—this was, in the final analysis, the Azeris' judgment on the matter. And as soon as the USSR—namely, Russia—supported Armenians, it was in turn nudged into the enemy camp.

In November, this evolution of opinion in Azerbaijan was reflected in the formation of a Popular Front. In a fairly conservative republic with little interest in perestroika, this event was highly significant. Azeris began a process that they claimed would make a radical change in the relations between the nations and the federation. The platform of the new Popular Front left few doubts about the role played by the Karabakh conflict in the fanning of nationalism.[15] At the heart of the platform was the question of Karabakh and the anti-Azeri actions of the central government through its biased information and meddling. But by way of putting the center on trial, this program also mentioned environmental issues and the problem of relations between Soviet Azerbaijan and Iranian Azerbaijan. It proposed, indeed, the opening of the border on certain days to allow for social and cultural exchanges between the two sides. The dream of a Greater Azerbaijan that terrified Lenin and Stalin gradually reemerged behind this new vision of national interests.

Another major source of discontent that set both parties against Moscow was the problem of refugees. Admitting refugees during a critical housing situation which the Armenian earthquake exacerbated, and feeding them while finding them jobs became an insoluble dilemma. Eventually a population of angry and bitter refugees in both republics became a disoriented mob with nothing to do and apt to join any movement expressing dissatisfaction. The number of "lost citizens"—refugees, formerly of Afghanistan—who felt they had been abandoned grew rapidly.

Neither Armenia nor Azerbaijan, where minds were already inflamed, needed desperadoes ready to side with those who promised them bread and jobs.

From Compromise to Civil War

Torn between Armenians and Azeris and uncertain of the policy that it should follow on the subject of the nations, the central government vacillated between one solution and another, from one side to the other. Similarly, faced with two adversaries that it could not appease, Moscow resorted alternately or simultaneously to the carrot and the stick. Its attempt to impose a compromise solution in early 1989 attests to a strategy in search of itself. For many reasons it was necessary to find a settlement acceptable to both sides and beyond that to the Soviet public. First, there was a mounting national crisis. Popular fronts were launched in most of the republics, and they followed the example of the Karabakh Committee—which decided the destiny of the nation on the spot and rallied the public around a plan. Moscow had to prove it was able to resolve the national conflict or it would lose all legitimacy at the periphery.

The urgency was great as elections for the Congress of the People's Deputies approached. The Soviet Union was by this time committed to a genuinely representative system[16] that mobilized the popular fronts. Could the whole field of national problems be left to these fronts, as an electoral argument? The stakes were high. Once again, the issue was to determine the direction that the people and the federation should take. Would Moscow or the capitals of the republics decide their futures? It was difficult to continue to ignore Armenian demands. The massacres and the earthquake had received international attention, and it was kept alive by a remarkably active diaspora. The world had great compassion for Armenia. With the leaders of the national movement in prison, it was important to find other negotiating partners in Armenia in order to avoid a mobilization of public opinion around "Armenian martyrs."

The quest for a compromise acceptable to both parties led Moscow to grant a special status to the bitterly disputed region. On January 12, 1989, Nagorno-Karabakh was provisionally removed from Azerbaijan's tutelage and placed under the authority of a special commission headed by a Russian, Arkadi Volsky, assisted by three Armenians and an Azeri. A kind of exceptional state was established; local authorities were temporarily suspended, and the commission dealt with all problems in cooperation with the neighboring republics. Moreover, the special zone was put under strict military surveillance. In a press release, the area commander stressed that all weapons would be recovered and that peace would be restored to a population that for a year had been accustomed to experiencing only demonstrations and strikes.[17]

Some months later, it became clear that all efforts to create conditions that would maintain calm had proven futile. In the spring, with elections over, there was further agitation in Stepanakert over the solution imposed by Moscow. Armenians may have suggested the temporary compromise of Moscow's taking the region in hand, but they sought a two-step solution allowing for Karabakh's annexation to Armenia to avoid a head-on clash with Azerbaijan. The special zone did not have this goal but was designed to cool tempers before a return to the previous status quo. Azerbaijan received assurances about the future stability of the borders, and Volsky attempted to make the closest possible link between Azeri authorities and improving living conditions in the area he presided over. Despite these appeasements, Azerbaijan resented the provisional reduction of its territory. They were even more reluctant to accept the commission's consultation with the Armenians. As for Armenia, it was outraged by a solution whose goal was not Armenian unity. All sides felt that once again Moscow's aim was to gain time without settling anything.

Despite problems in Armenian elections (the Karabakh Committee, the true representative of the people, was outlawed and could not participate, and violent incidents paralyzed the balloting), the elections of March 1989 made it possible for the Armenian deputies to the congress to renew the debate. They proposed a referendum giving the population of Karabakh full exercise of the right to self-determination. Although referendums were provided for in constitutional documents, in practice they never occurred in Soviet politics. In the complicated ethnic context of the Soviet Union, allowing a referendum to make it possible for a national group to determine its own borders or level of participation in the federation would amount to opening the floodgates to a spate of demands for referendums on self-determination.

The hostility of the republics containing separatist communities within their borders considerably expedited Gorbachev's rejection of this demand.[18] For once, he could rely on a broad national consensus against a national project. To influence Gorbachev's decision, the deputies from

Azerbaijan used an argument that was repeated a year later by Moscow in its conflict with the Baltic countries when the latter in turn demanded their independence. For the Azeris, any referendum had to be republic-wide; thus, the entire population had to address the demands of the Armenian minority. The majority position can easily be surmised. The course of this debate in the newly elected congress accurately reflected the mood in the area: 5,500 men from the armed forces held less and less of Karabakh, where clashes between Armenians and Azeris led to daily deaths and injuries.

Suddenly, the situation shifted. Gunshots and sporadic attacks gave way to a major mobilization in both republics when transportation was blocked. First the Armenians isolated Nakhichevan from Azerbaijan by sabotaging the railway lines. The Azeris responded by blocking the railway system to Armenia, depriving that country of food, gasoline, and construction materials for repairing the earthquake damage.[19] The escalation included a rail blockade, attacks on trains, the mining of bridges, strikes that paralyzed the autonomous region, pitched battles between communities, acts of sabotage, abductions, attacks on military posts, and an appeal in Stepanakert and everywhere Armenians and Azeris lived together for the organization of armed self-defense groups. In the fall of 1989 civil war broke out in the Transcaucasus.[20] It even threatened to spill over into Georgia, which Azerbaijan demanded should associate itself with the rail blockade on pain of itself being blockaded.

Moscow had a flood of reactions to this situation, which was new to the Soviet Union. First, Gorbachev put Azerbaijan on formal notice to restore transport, which was ignored. On October 3, a week after this ultimatum, the Supreme Soviet of the USSR did the same, with the same lack of results. Clearly, the central government had lost all authority over the Transcaucasus.

A similar contempt for injunctions from Moscow may be found in the decisions by those elected as part of the region's special status decided in January. In August 1989, a congress of representatives of the population of Nagorno-Karabakh—supported by the leaders of the Karabakh Committee, who had earlier been released—met in Stepanakert, decided to abolish this special status, proclaimed the independence of the territory in the framework of the USSR, and placed it under the authority of a National Council. Throughout all this, no one even informed Moscow about what constituted a territorial and political revolution. The Supreme Soviet of Azerbaijan did demand that Moscow disband the special commission and return Karabakh to its authority. Confusion was rampant.

In Armenia, the three colors of independence waved over all the demonstrations, and people sang the national anthem. The Karabakh Committee sent deputies to the congress in Moscow and expanded into the National Panarmenian Movement.[21]

Civil war combined with political war to expose the powerlessness of the central government; it was unable to disband a serious blockade that one republic organized against another;[22] unable to enforce the special status decided on in January, which some wished to eliminate, others disputed, and still others ignored; unable to halt the escalating violence, while tens of thousands of men—troops of the MVD and the regular army—were assigned to the region. In the fall of 1989, the Transcaucasus slipped out of Moscow's grip and landed in a world ruled by local passions.

The central government's paralysis was all the more striking because a few months earlier, Moscow had tried to check national aspirations by repressing demonstrations in Tbilisi. The meaning of this repression was well understood, but was rejected—first in the Transcaucasus, and later elsewhere, it was evidently decided that, repression or not, Moscow could not and must not have a say in these matters.

Lebanonization

In September, the president of the Special Commission, Arkadi Volsky, acknowledged the failure of the Moscow-imposed compromise in Nagorno-Karabakh. In a disintegrating country paralyzed by strikes and blockades, the point of maintaining a system that no one agreed to submit to was doubtful. Above all, he observed that the only authorities able to make themselves heard were those coming from the people. Why then persist in systems that made a mockery of the central government?

On September 28, the Supreme Soviet of the USSR agreed, and terminated the authority set up less than a year earlier. This was a wise move, no doubt, but one that scarcely helped bolster Moscow's prestige in the area and also augured ill for the chances for the success of the new system advocated by the Soviet.

The region's special status was replaced by a complicated procedure that was bound to fuel the conflict. Karabakh was placed under the authority of Azerbaijan, which, however, had to take appropriate legislative measures "to guarantee the region's autonomy" and maintain public order and security. A new commission, mandated by the Supreme Soviet of the USSR, was to watch over it,[23] while the troops of the MVD—from 5,000 to 6,000 men—maintained order. Power in the region came from two authorities: the regional Soviet, deactivated a year before, and a Mixed Organization Committee in which Armenians and Azeris had to be represented on the basis of their numerical weight in the region (three-quarters Armenians, one-quarter Azeris). Azerbaijan was told not to try to change the demographic balance by shifting the population (by resettling colonists or terrorizing the Armenians to get them to flee), as it had done in the past.

Once again, the reactions of the interested parties confirmed that Moscow had not produced the miracle solution. On December 1, the Supreme Soviet of Armenia declared that Karabakh was part of the United Republic of Armenia and refused to recognize the new Soviet decree. There was fury in Azerbaijan; Azeris felt that the republic's authority over the region had not been affirmed in sufficiently clear terms, and they did not accept Moscow's unilateral decision. Against what they took for an intolerable encroachment on the sovereignty of the Azeri state and a challenge to its authority over its own territory, the government of Azerbaijan—Soviet and party—and the Popular Front gave the signal for popular demonstrations in Baku and the rejection by the Supreme Soviet of the republic of the decree of September 28, for unconstitutionality.[24] The Azeri government created its own committee and, to administer Karabakh, put it under the sole control of the republic's authorities. The standoff was complete. These opposing decisions were humiliating for the Soviet government, whose authority was held up to ridicule more than ever before. Apparently the parties felt that the only solution to the conflict in a region in which the deaths could no longer be counted was force, including armed force.

For Gorbachev, 1990 started off both well and badly. In the West, he was "man of the year" and spoken of as a future Nobel Peace Prize laureate. In the Transcaucasus, however, disaster was accelerating. As soon as the new status was adopted in Moscow and rejected in Yerevan and Baku, the two republics—for from then on this was a direct clash between the republics—were launched on a political auction and an escalation of violent incidents that further undermined the central authority.

The escalation was at first political. The Supreme Soviet of Armenia registered as a fact the incorporation of Karabakh. The Armenian budget treated this region like the rest of the republic; administrative measures for the region were decided in Yerevan. To put an end to the injunctions and reprimands from Moscow, the Armenian leaders proclaimed on February 11 that federal laws would apply in the republic only with their consent and could be vetoed. This decision, already made in other republics and declared unconstitutional by the Supreme Soviet of the USSR in November 1989, is evidence of the revolution the Karabakh affair had wrought in Armenia. In less than a year, the republic most in need of Soviet protection was in a state of complete rebellion. Fear of the Turks had been overcome; it was in any case weaker than the scorn henceforth exhibited for the central government and the federal order.

The Azeris, for their part, responded to Moscow with violence and blackmail. Armed bands controlled by the Popular Front seized all the public buildings in the Azeri town of Lenkoran where the government of the Front was established. From there, they thumbed their noses at Moscow and announced that the city would remain out of Soviet hands as long as Nagorno-Karabakh was not returned to Azerbaijan. Taking a whole town hostage was certainly a brand new way to negotiate. But the violence did not stop there. Troops were needed to guard the trains that were inspected. One after another, bridges were blown up, and barracks ransacked to supply combatants with precious arms. People fought in the cities and in villages. Helicopters of unknown origin fired at Azeri towns; pogroms organized in Baku ravaged what was left of the Armenian community. Disorder also prevailed at the borders, notably in Nakhichevan, contiguous with Iran, where the Azeris, dismantling the border installations, intended to indicate their solidarity with the rest of the republic.

The Soviet government finally recognized its failure. "It is civil war," said the minister of the interior Vadim Bakatin.[25] The official toll of the victims of the pogrom of January 13 was more than sixty dead and was no doubt more in reality. The fighting, still raging in Stepanakert, reached Kirovabad, Azerbaijan's second-largest city. It was not until January 18 that the minister of defense, Dmitri Yazov, announced that the regular army would reinforce the MVD to reestablish order; a state of emergency had already been announced three days earlier,[26] and the forces of the MVD had been authorized to shoot on sight to protect themselves and above all to defend the arms depots. In Baku, 17,000 men from the security forces patrolled, reservists were called up, and the military units were reinforced. On January 19, an attack was launched on Baku, where the army tanks had trouble clearing a path. Popular resistance was organized and forced troops to treat the Azeris as combatants in a regular conflict. Azeris blockaded the port, which the army opened after bitter fighting. The dead on both sides, the troops and the Azeri resistance, numbered in the dozens. The reestablishment of order—more than 30,000 soldiers had been sent to Baku to reinforce the MVD—was long and precarious. A curfew and the prohibition of all demonstrations and strikes did not prevent the country from being paralyzed by a quasi-general work stoppage and outbursts of violence, and transportation remained partly cut by sporadic blockades. The population of Azerbaijan did not accept its subjection to military force; three weeks after the expedition, the Soviet press, which was more candid than the government, admitted that more than half of businesses and workers passively resisted the army's authority.[27]

To civil war was thus added regular war and organized resistance to Moscow. The Transcaucasus was quickly being Lebanonized. Confronted with this dramatic development, the central government was faced with finding an effective solution. Its use of armed force may have risked creating a major conflict in the area and perhaps was insufficient to restore its own authority.

To justify this war in Azerbaijan—for it was very much a war, even if it seemed like a blitzkrieg—Moscow accused the Azeri nationalists of being on the brink of achieving three goals: the overthrow of the Soviet government, the secession of Azerbaijan, and the formation of a unified Islamic Azeri state.[28] Gorbachev clearly declared that military intervention was inevitable to restore order and "bar the way to the conspiracy to hand the government over to extremists." Restoring order would save Armenian lives threatened by the rise of violence. In the face of widespread outrage—the Armenians themselves sharply criticized the use of armed force—this humanitarian argument found increasing favor, which led to an initial question: Why was such heavy use made of armed force? What really was achieved by converting Baku into a battlefield?

One response might be a desire to save human lives in a climate of rampant civil war. The decree imposing a state of emergency on January 19 indeed spoke of "ensuring the citizens' safety." It was true that there was no safety, and that the intervention occurred when the pogrom of January 13 had already produced dozens of victims. A careful reconstruction of these events shows that this pogrom was not only foreseeable, however, but that local authorities were not in a position to be unaware of its organization and nothing was done in time to prevent it. It had served only as a pretext for the intervention.[29] The authorities' dealings with the most radical Azeri elements were bound to fuel this argument. The anti-Armenian demonstrations of 1988 that ended in the Sumgait pogrom had been in large part organized by a remarkable ringleader, Nemat Panakhov, turner in a Baku factory and an ill-educated man but one with a gift for agitation. Panakhov, called "a son of the people," was the party leader of the demonstrators and the one who made the most hostile demands regarding the Armenians. He also gave the demonstrations a strongly Islamic turn—green flags and portraits of the ayatollah Khomeini in the marches—which later were forbidden.

A year later, after many incidents, Panakhov, freed from prison, was expelled from the Popular Front, which he accused of compromising with the Armenians. It was at that moment, on January 12, that Azeri television offered him air time for speaking to his compatriots who had fled Karabakh. The message was that, although they had no housing and were in need, they would soon see evidence that the state was investing heavily in the Armenian enclave in order to make up for economic disparities. Thus the most radical and violent of the leaders of the Azeri movement was given the means to fan a fire that was worsening every day. It was not surprising that a pogrom began the following day.

After the intervention, the first secretary of the local Communist party, Vezirov, was dismissed from his post. It was necessary to act ruthlessly. But it is noteworthy that the second secretary, the Russian Polianichkov, who had "normalized" Afghanistan during the war and whose strange activities in the Azeri nationalist circles had been noted by observers, escaped dismissal and even criticism. There was no doubt, however, about his responsibility for the agitation and excesses. Evidently the pogrom had been useful to Moscow in that it allowed the government to send into Azerbaijan a veritable expeditionary force that first proceeded to break up the national movement. Did the local, and then the central, authorities simply close their eyes to the coming pogrom? Or did they hasten the violence to set a trap for the Azeri nationalists? Whichever rationale they used, the military operation tried to neutralize the most radical Azeris and not to preserve Armenian lives that had already been sacrificed.

Soviet Power in Question

Although in 1990 the central government was more hostile toward Azerbaijan that it was immediately after the pogroms of 1988, and although it seemed more inclined to favor the Armenian demands, this was less because of a subtle seesawing than because it suddenly appreciated public opinion in Azerbaijan. More belatedly than the Armenians, but in the same ways—the mobilizing action of a popular front and related organizations around the regional conflict—the Azeris eventually felt that they had to settle their dispute without Moscow's assistance and even against Moscow's wishes.

Three organizations were the source of the crystallizing anti-Soviet Azeri movement. Chief among them, when the political movements had not yet any legal foundations, was the Shenibel cultural association,[30] which was concerned with revivifying the Azeris' historical and architectural heritage and with safeguarding their language. In 1987, the association received the support of a strange and dubious Azeri character from Iran, where he had fought in the ranks of the Fedayeen before fleeing the Khomeini regime to the USSR. This Irano-Azeri, Siyamet, together with the historian Hatemi, trained and guided Panakhov, the turner from Baku whom the anti-Armenian demonstrators of 1988 brought into the spotlight. The Panakhov-Hatemi-Siyamet trio can be compared to the Polish group formed by Lech Walesa, Bronislaw Geremek, and Adam Michnik: the worker figurehead of the movement and the intellectuals who attempted to channel the popular energy that this leader could mobilize. The three men were arrested after the riots and their movement dissolved, but a few months later, the local authorities freed those who were Azeri citizens. Meanwhile, Shenibel, stripped of its leaders, returned to its first business, preserving the national culture, and depoliticized itself.

Hatemi and Panakhov, who in the fighting and then in prison had developed a radical conception of the struggle, turned momentarily to other organizations. Their range of choices was limited. The oldest one, Birlik (Unity), which numbered the historian Aliev among its leaders, leaned toward Islamic nationalism. Their watchword, "unity," referred to the formation of a great Azeri state uniting north and south. The underground contacts between the two parts of this nation naturally resulted in opening Soviet Azerbaijan to propaganda from Iranian Islam.

The religious extremism of the Birlik association next favored the

Azeri Popular Front, founded at the beginning of 1989 by Azeri intellectuals who wanted to keep the crowds eager for demonstrations away from the extremists, religious or not. Their model was the legalistic popular fronts of the Baltic states that tried to involve local governments in national demands and wished to have a dialogue with Moscow within the framework of the unity thus achieved. Like the Baltic fronts, the Azeri Front made an attempt in the spring of 1989 to reach an agreement with the Communist party, but the attempt failed. Vezirov, first secretary of the Azeri Communist party, obviously feared that the Popular Front could rally the whole population around the anti-Armenian theme, whereas Moscow's hesitant policy was paralyzing the local party and the whole population. The party refused to envisage a competition in which the front would move beyond its extremism and therefore treated it as an enemy that threatened the whole system.

This undisguised hostility, which resulted in repeated skirmishes, did not prevent the front from getting organized within a few months, establishing local authorities in many communities throughout the country, and presenting a detailed program not only to its sympathizers but also to other popular fronts. To the classic categories enumerated in these documents (economic and cultural renewal, autonomy, human rights, ecology) were added national arrangements that were immediately relevant for this population: sovereignty for Azerbaijan (which included the problem of Karabakh) and the membership for the republic in international organizations; the merging of the two Azerbaijans; unifying the republic through the Azeri language, which was given the status of the only official language. With strong popular support, measured by the successful demonstrations it had organized, the Popular Front demanded that the Party recognize it and allow it to run candidates in the elections.

Nevertheless, because it had been unable to take part in the elections in the spring of 1989, the front cast doubt on their pluralistic nature and demanded a cancellation of the vote and the election of new delegates to the Congress of the People's Deputies of the USSR. Through its demonstrations, it called for the freeing of the imprisoned national leaders of Birlik and Shenibel. It continued to exert pressure on the local authorities.[31] The Popular Front's position remained moderate up to the middle of 1989. But the government worried about its ability to foment strikes, notably among railroad workers (who organized the railway blockade)[32] and to marshal ever more numerous crowds in Baku and other cities. It was also worried about the rapidly growing legitimacy that the front enjoyed with the people and among the popular fronts in the other republics, which were gradually increasing their coordination with each other. Eventually the Communist party and the government of Azerbaijan were provoked into acting by the rise of Islam, which the front unsuccessfully tried to impede, and by the radicalization of leaders like Panakhov, who first rejoined the front, tried to make it a real tool of war, turned its nonviolent demonstrations toward violence, and finally left it while accusing it of colluding with the government and the Armenians (which, however, forced the front to harden).

At the end of 1989, aware of the real danger that the front would steal all its support among a population that was ever more inflamed over the Armenian question, the local government drew up a gloomy estimate that was no doubt accepted in Moscow. The Azeri party was discredited for both its "pro-Moscow" attitude and old habits of corruption of which the Popular Front accused it. The front was becoming increasingly radical as were the people. As between an increasingly radical populace and an increasing radical front, it was impossible to decide which was pushing which. But the dynamics of revolution demonstrate that there comes a moment when no one can interrupt such a linkage. The result was that the Popular Front set itself up as the true representative of the society in opposition to a discredited government. Moscow's vacillation—siding with the Armenians with special status for Karabakh, and then with the Azeris with the abolition of this status—exacerbated this radicalization, for sometimes this hesitant policy gave Armenians hope and enraged Azeris, and sometimes it did the reverse. Ultimately, the two adversaries simultaneously lapsed into frustration and mistrust.

In January 1990, the local government felt besieged by the crowd and overwhelmed by the Popular Front, which went from one demonstration to another, daily increasing its demands. If the local government and Moscow refused to yield to these demands, then it was necessary to break with both local authorities and Moscow. The time for a popular, national revolution against the Soviet government appeared to have arrived.

Gorbachev's options were limited: he could either (1) restore his authority and that of his representatives by a show of force, which implied exercising violence, outlawing the nationalist authorities, and minimizing the popular resistance—in short, a real war; or (2) yield to national demands, which meant totally sacrificing Armenia, allowing an unprecedented Armenian mobilization, and assuring all the demonstrators and popular fronts of the USSR that their fight could succeed in cowing Moscow. Perhaps, at the end of the road, defeat was inevitable.

Gorbachev chose the first solution—a war against Azerbaijan to save the Soviet government. The price of this was the Lebanonization of the entire region.

In the two years from 1988 to 1990, a territorial dispute that for decades simmered at the stage of a complex but manageable problem, transformed ethnic relations in the Transcaucasus and the relations between the people of the region and Moscow in a way that was hard to reverse. Even though Armenians and Azeris continued fighting, they understood that they had to hold a dialogue and did so within the national framework, without Moscow's mediation or arbitration. By the end of January 1990, the first face-to-face encounter took place in Georgia and then another in Riga. The Georgian Democratic National Party argued before the Armenian and Azeri fronts for the necessity of a truce and an arrangement ensuring peace in the Transcaucasus. Progress was slow; it could lead—at least in the foreseeable future—to more concessions needed for settling urgent problems, such as the fate of the prisoners and the wounded, than to a true honorable and lasting peace. The wounds were so deep that much time would be needed before the lives of the varied people of the Caucasus could again return to normal. All were agreed on one point, however: hostility was henceforth to be directed against Moscow.

Following the Baku massacre, the two parties lost little time in noting that in the decisions made at the center, political calculation took precedence over humanitarian concerns. The state of emergency in Azerbaijan was announced too late to save anyone. In Karabakh, where at that moment there was less urgency, the announcement occurred on January 15 and served primarily to justify the arrest of all the leaders of the Armenian National Council. Aware of this casual attitude toward their lives, Armenians responded with a provocation. Suddenly, they showed solidarity on this point with the Azeris and denounced the war being waged on them. After this war, in every demonstration in the Caucasus, Gorbachev's name was booed and larded with abusive epithets, of which the most common were "murderer" and "fascist." Armenians and Azeris agreed that their situation, at first poorly understood in Moscow, had later been inflamed and used to maintain the cohesion of the empire. But they refused to blame the KGB or the army for what they took to be a crime done to particular groups. Quite the contrary, they noted that the army found it difficult to carry out military expeditions on Soviet territory. The call-up of reservists for the operation of January 20, which had led to many examples of refusal to serve, had convinced them that it was not the army but the system and those who directed it that were fully responsible for the choices made.

The Transcaucasus's mobilization against the Soviet system was not just political but also military. The existence in the two republics of true national military units, well-equipped with stolen or purchased weapons, was an established fact. The regular Soviet army had had to learn how to deal with ambushes and attacks on its own territory, which considerably worsened its morale. In two years, the local problem of Nagorno-Karabakh led to the transformation of the whole Caucasus into a powder keg that threatened the Soviet system in its entirety.[33]

5

Black Sunday

ON Sunday, April 9, 1989, Tbilisi, the capital of Georgia, was the scene of one of the bloodiest clashes in the Gorbachev era. Troops dispersed a nonviolent crowd of demonstrators who had gathered in support of hunger strikers demanding independence for the republic. Soldiers armed with shovels indiscriminately struck men, women, teenagers, and children who were massed on a square that was hard to flee from and who thus found it hard to obey orders to disband. The official toll of the violence—more than twenty dead, some 200 wounded—or the much heavier one reported by the demonstrators does not reflect the consequences of the episode. In the collective consciousness of the Georgians, April 9, 1989, became a symbolic date harking back to February 25, 1921 (when Georgia was annexed by Soviet Russia) and August 24, 1927 (when a Georgian uprising was bloodily crushed) in a long list of conflicts and hatreds that fed an increasingly impassioned desire to sever the ties between Russians and Georgians forever.

Within the dynamic context of Gorbachevism, the tragedy of April 9 gave a conclusive impetus to the Georgian national movement, which had been exuberant in its expression but long hesitant about its concrete demands. Like the crisis of Nagorno-Karabakh, this tragedy disclosed the complexity of the national question in the USSR, ambiguities about the desires of each group concerned, and the central government's inability to come up with a coherent response to the conflicting demands. The date was tragic for Georgia but also was perhaps one of the most decisive events marking the end of the Soviet empire.

The Abkhazian Imbroglio

The cause of the massacre of April 9 was the problem of Abkhazia. The demonstrations in Tbilisi had been sparked by a seemingly local but long-standing conflict between Abkhazians and Georgians. Abkhazians complained that Georgia had forgotten them and demanded the right to secede from it and rejoin the Russian republic; Georgians accused Moscow of systematically exacerbating the conflict and using it as a pretext to intervene in Georgia and thus reduce the possibilities for national independence. These were two conflicting versions of what at first seemed a highly circumscribed problem and parochial quarrel in which Moscow could comfortably play the arbitrator. In reality they concealed a mine that was set off by Moscow and in 1990 complicated the specifics of Soviet domestic policy.

Some geographical and historical facts clarify the issues. The borders of the Republic of Georgia include the Autonomous Republic of Abkhazia. Georgians far outnumbered Abkhazians: according to the 1989 census, there were 3,983,000 Georgians in the USSR (as contrasted with 3,570,000 in 1979) and only 102,923 Abkhazians (as against 80,915 in 1979). The Republic of Georgia had 3,789,385 Georgians and 93,275 Abkhazians.[1] A minority in Georgia, the Abkhazians were also a minority in the autonomous republic, where they numbered 90,213, while the 242,304 Georgians were almost three times as numerous. This hefty Georgian presence in the autonomous republic was a source of apprehension for the Abkhazians. Between the two most recent censuses—1979 and 1989—the number of Georgians in the area grew more quickly than that of Abkhazians, even though the birth rate of Abkhazians was slightly higher than that of Georgians. Clearly, a population movement by Georgians into the small Abkhazian republic was underway. Moreover, Georgians were totally cut off from the cultural life of the republic since only 0.3 percent of Georgians knew the Abkhaz language. Their refusal to integrate with the republic was unmistakable.[2]

The problem was not new. Like that of Nagorno-Karabakh, it went back to the beginning of the Soviet federation and followed the same logic. In the 1920s, Stalin (and Lenin had never disavowed him on this score) wished to stifle Georgian nationalism by employing the minorities populating the republic, primarily the Abkhazians. Since the nineteenth century, this Muslim people had rebelled against two Christian influences—the Georgians and the Russians. When the czarist empire annexed Georgia in the nineteenth century, many Abkhazians fled to neighboring Turkey to escape a Christianization that they rejected but that in fact never took place. After the revolution, faced with an independent Georgia and then with Georgian national resistance, Stalin decided to play the Abkhazian card by creating a territory for Abkhazians within an autonomous republic. It scarcely mattered that they represented barely a third of the population and that the Georgians, who were in the majority in the area, fiercely opposed this solution. Trusting in the old principle of divide and conquer, Stalin proclaimed that in the territory bearing its name this minority would have preference over other ethnic groups, notably the Georgian majority.

So the conflict had existed since the early 1920s. For the Georgians, Abkhazians were a tool in the hands of Moscow, which was untrustworthy and would eventually have to be broken. They also formed a tight enclave within the Georgian territory with no tolerance for the political and cultural rights of Georgians. Since Georgia was too small and vulnerable to form a federal state, the Georgians appealed to history to show that their country could be a land of hospitality and tolerance as long as it remained Georgia, not a conglomerate of conflicting desires and identities. Conversely, the Abkhazians tried to consolidate their republic by staying separate from the dominant Georgians. The Abkhazians dreamed of and asked the Soviet government for the return to their state of the Abkhazians who had settled in Turkey in the nineteenth century and to give them back the lands "wrongfully occupied"[3] by the foreign colonizers (that is, the Georgians). A people beloved by Stalin in the 1920s, the Abkhazians experienced a period of disgrace during the first five-year plan and even more after World War II, when their former protector even thought of adding them to the list of peoples to be deported for collaborating with the German army.[4]

For more than thirty years, the Soviet government waited for the national problem to disappear. The little local nationalisms, such as that of the Abkhazians, or the larger nationalisms, such as that of the Georgians, had been transcended by government decree; therefore, any problem was simply a matter of regional frictions and antagonisms. In 1978, however, it became clear that the USSR still had a national problem and that it was particularly acute where Stalin had once tried to eradicate it by exploiting the hostilities handed down by history.

The Constitution of 1977 caused all the hitherto creeping antagonisms to erupt. Possibly because the Soviet government emphasized the existence of a new historical stage—that of an emerging Soviet people—and perhaps also because it threatened a radical integration of the nations within a unitary state, national aspirations were being affirmed.

The Georgians held mass demonstrations on behalf of their own constitution, adopted in 1978, to recognize Georgian as their official language. What seemed self-evident in fact amounted to a revolution. It meant that in a minifederated Georgia (including two autonomous republics and an autonomous region), there would be a diminution of the rights of the minorities with self-governing status: Abkhazia, Adzharistan, and Ossetia. In the face of the Georgians fighting to be culturally dominant within their state, the Abkhazians then continued their fight in the economic arena, claiming that Georgia was keeping them materially as well as culturally underdeveloped in order to obliterate the Abkhazian identity. In 1978, under this threat, they demanded immediate annexation to the republic of Russia, preferring to be a minority in the great multiethnic federation than to be in a small state bent on absorbing some minorities in an attempt to face off against the Russian Big Brother.

Georgia, of course, could not accept this draining of its population and reduction of its territory. As early as 1978, the Georgians thought that behind the demands of the Abkhazians they detected some manipulation by the central government. Faced with their violent reaction, Moscow capitulated: in 1978, the USSR was unprepared for popular demonstrations. Better to offend 100,000 Abkhazians than 3 million Georgians. Let them rally around their language and culture. By way of compensation, the central government obtained from Georgian authorities a plan for economic and cultural development in Abkhazia.

Suspicions did not go away on either side, however. The Georgians settled colonists in Abkhazian territory to reduce the Abkhazian proportion of the local population. The 1989 census confirms this. The Abkhazians worried about this increase in the Georgian population and endeavored to make their lives difficult. Despite the lack of precise information from before 1985, it was clear that interethnic relations in Abkhazia were continuing to deteriorate and that the daily lot of members of both adversarial communities were humiliation and violence.

Abkhazian local police were rumored to display systematic leniency when the victims were Georgians. The Georgians claimed that "ethnic crimes" were a common practice, and that they were treated as pariahs and barred from jobs and housing. Because of the unreliability of the sources and the emotional climate surrounding this history, it is too soon to decide who was the oppressed and who the oppressor in Abkhazia in the 1980s, but during this time a muffled hostility was transformed into overt hatred. Glasnost allowed the disagreements to be openly expressed, which ultimately caused the hatred to spill out into the public arena. Still, glasnost did not escalate interethnic tensions; it simply made it possible to take its measure.

From Escalation to Massacre[5]

In 1989, while violence flared over the problem of Nagorno-Karabakh, the potential explosiveness of small nationalities in the sensitive national climate in the USSR frightened the ethnic groups that were attached to other groups against their will. Georgia feared that Moscow would later give in to the Abkhazian pressure for secession as soon as the Armenia-Azerbaijan conflict raised the general problem of redrawing the internal borders of the USSR and redefining federalism. In Tbilisi it was also feared that local violence would provoke Moscow to make an example of Georgia: in this case, the absence of an influential diaspora would help avoid the negative external consequences of a show of force. Furthermore, the Georgians suspected that the central government was repeating Stalin's game—that is, in a period of national agitation, trying to provoke a crisis among them (supporting Abkhazians to allow a show of force) that would serve as an inexpensive warning to other restless nationalities. All these reasons combined so that hatred of the Abkhazians reached a paroxysm.

Conversely, Abkhazians were more intransigent than ever in demanding separation. Fueling their nationalism were concessions that Georgia was forced to make after 1978. Now instability of a USSR dominated by strong national feelings led them to seize this time as propitious for stating demands that Moscow would have to listen to.

Thus opened an era of all-out hostilities. In June 1988, in a document addressed to the Twenty-ninth Party Congress, the Abkhazians demanded the right to secede from the federation. They then fired off petitions on this theme to all the decision-making centers, party and state, in the USSR and Georgia. On February 18, 1989, several thousand people marched in Tbilisi to protest the idea of a secession by the Abkhazians, whom they accused of sabotaging interethnic relations by keeping the Georgians out of all leadership positions in Abkhazia.[6] "No to secession!" "An end to discrimination!": these two slogans summarized the whole way of thinking underlying the activities of the February demonstrators. They attacked on two fronts—Abkhazian behavior in Georgia and Abkhazians' complicitous relations with Moscow—and drew a clear conclusion: a minority that was so small and so dangerous for Georgia's future must be given decision-making authority based on its population figures.

Responding to this thinly veiled threat about their future status, the Abkhazians broadened their demands and expressing them more emphatically. On March 15, 1989, the Abkhazian village of Likhmy, where the secessionist movement had started ten years earlier, was the center for the largest popular mobilization ever seen in this area. Not satisfied with asking to secede from Georgia, the Abkhazians added a new demand—an independent status giving them equality with the Georgians, the status of a sovereign republic. This demand was inconsistent with the Soviet Union's rules for forming sovereign republics, whose minimum population must be 1 million. In 1989, Abkhazia's total population was only 524,000, of whom the Abkhazians were less than a fifth. But, as whimsical as it was, this demand confirmed that the Abkhazians were determined to leave the republic of Georgia by any means. At the time the central government in Moscow proclaimed that, faced with rising national tensions, its duty was to preserve the rights of *all* minorities, whatever their numbers and location. No one could say that the demands of the Abkhazians were totally hopeless.

Tiananmen in Tbilisi

These conditions led to an urgent desire in Tbilisi to make an explosive display of hostility and to show their capacity for resisting this plan. Beyond the local conflict that many used as a pretext, however, the main purpose of the demonstrations of February 1989—notably that of February 15—was to recall Georgia's forcible annexation to the USSR in 1921. Between February 1988 and February 1989, the annexation was commemorated in a way that revealed how extensive the change in political climate had been. On February 25, 1988, there was no public gathering, but sadness and solemnity permeated the streets of Tbilisi. A year later, the crowd booed the annexation of 1921 and demanded independence. The local government panicked and spoke of "destabilization" and of a struggle for the "conquest of power."[7]

On April 4, 1989, a crowd estimated at 20,000 men and women of all ages overran the square and neighboring streets around the building of the Council of Ministers. For five days, this peaceful but determined crowd jubilantly chanted slogans that were evidence of a rapidly evolving movement. In a few days, the crowd quickly grew from the initial 20,000 demonstrators to more than 100,000, and a quasi-general strike paralyzed all public services. During the crisis, an atmosphere of insurrection was clearly taking over even though the crowd remained good-humored. Certainly, hostility toward the Abkhazians was present, but the popular will brought to the forefront what thereafter inspired all the demonstrations: Georgia's right to decide its own destiny and resolve its own problems. Until then, the speeches in the popular assemblies had been confined to the theme of greater autonomy. In April 1989, the Georgian demands shifted from autonomy to independence. On April 4, some militants claiming to represent a still-illegal party, the Georgian National Democratic Party, attempted to obtain their homeland's independence by going on a hunger strike across the street from the seat of government. They said that all details had to be resolved within the framework of their independence. At first diffident about this new theme, the crowd gradually took it up and demanded more and more forcefully that the first stage was the departure of the Russians from Georgia. In fact, this had already been underway for nearly thirty years. In 1959, 408,500 Russians were living in Georgia; in 1970, 397,000; in 1979, 371,000; and in 1989, only 338,000, while during these latter decades, the total population of the republic went from 4,044,500 to 5,395,841 inhabitants.[8] Popular hostility had long encouraged the Russians to leave, and in the last decade this movement had accelerated. For the first time, however, the statistics were transformed into slogans, and Moscow was certain to be frightened by them.

It appeared that if the movement remained nonviolent, it was destined to widen. On April 8, a large demonstration in Kutais showed that the call for independence was igniting the republic. And it was clear that the local authorities were unable to control the situation. That same day, the first secretary of the Communist party, Dzhumbar Patiashvili, exhorted the demonstrators to remain calm. Where Eduard Shevardnadze had in 1978 successfully argued that he could play the role of mediator, Patiashvili failed to make himself heard. The crowd could march and express itself with impunity. And in the USSR of 1989, the authority of the representatives of the Party was increasingly precarious. Deaf to Patiashvili's appeals, the demonstrators indicated their desire to go forward with their movement toward the independence they believed was in sight. In Sukhumi, the capital of Abkhazia, the Georgian population echoed the demonstrators in Tbilisi by loudly expressing their opposition to the Abkhazians' demands for secession. By the morning of April 9, Georgia was paralyzed by three conflicts: one between the republic's population and the Soviet federation, another between Georgians of the autonomous republic and the Abkhazians, and yet another between the Abkhazians and the Georgians. A conflict had occurred between the periphery and the center, and interethnic and intercultural conflicts were occurring at the periphery. All called on the "Russians" either to grant Georgia its destiny or to admit Abkhazia into its borders. This combination of conflicts and contending desires created a difficult situation and underscored both the central and the local authorities' inability to come up with any mediation or quick conciliatory response.

Instead, the leaders decided on repression. The result was the massacre of black Sunday, whose human losses and the responsibility for it would immediately be the subject of impassioned debate. In fact, at dawn on April 10, a battle of numbers began; were there from sixteen to eighteen dead, as the authorities stated,[9] or hundreds, as the Popular Front, organizer of the demonstrations, maintained?[10] All versions concurred about overburdened hospitals taking in the wounded. Soon, however, the government tried to deny—before they fell silent about it—a dramatic fact of the repression: paralyzing nerve gasses of unknown composition were used against the demonstrators. This was the first accusation of this kind against the government, but a short time later, these gasses were used in Moldavia, which had hitherto been spared national clashes. The official response was that the demonstrators had not been gassed but had suffocated in the crowd movements. The investigating committee would bring these methods of repression to light.

Another debate concerned the demonstrators' goals. The first news that circulated in the USSR accused the nationalist organizations of two crimes: they were spreading interethnic hatred and thus threatening the physical safety of the Abkhazian community, and they had also tried to organize secession and set up a provisional independent government. But these accusations, made in Moscow on April 10, did not correspond to reactions on the spot. The republic closed its borders to journalists, and, immediately after the tragedy, the local government tried to reestablish contact with the shaken population. Appearing on television, the first secretary of the Communist party, Patiashvili, deplored the "national calamity" and suggested that there might have been some provocation. On April 12, before the republic's Politburo was hastily convened, he proposed to resign, and by April 14, he was replaced by Zhivi Gumbaridze, an apparatchik who had for some months been president of the KGB. Disorder had to be met with the naming of a specialist in the maintenance of order. All the local party and state leaders—president of the Supreme Soviet, head of the government—resigned from their posts or offered to do so, reflecting a situation difficult to explain and justify.

Moscow's reaction was just as noticeably anxious. Contrary to predictions, the violence of the repression did not shatter the Georgian people. On the morning of the tenth, demonstrations in favor of independence were replaced by demonstrations of solidarity with the victims. The next day was proclaimed a day of mourning by both the government and the Popular Front: a short time later, the people of Tbilisi gave their dead funerals that were calm, but the huge crowd created the effect of a demonstration.

In the local government's state of disarray, it was the man of 1978, Eduard Shevardnadze, who seemed most capable of resuming the dialogue between his compatriots and the Soviet system. Canceling all engagements as an active minister for foreign affairs, Shevardnadze came to Tbilisi, accompanied by Razumovsky, another member of the Politburo.[11] Of the two of them, however, Shevardnadze was the real player in this attempt at reconciliation. He had discussions with the local authorities and leaders of the national movement, for just after April 9, the people recognized no other representatives than these groups. He tried to bring the two conflicting positions and demands together, telling the political leaders that the killing of peaceful demonstrators was unacceptable—which was also a way to exculpate the central government, whose spokesperson he was, of the tragedy of Tbilisi. He told the intellectuals and representatives of the Popular Front that the repression had not been an innocent choice but a dirty trick played on perestroika. Thus the official thesis appeared: on the one side were those who opposed the renewal of the USSR and who, in the mounting disorder in Tbilisi, had found fertile ground for turning the people against perestroika; and on the other were the supporters of perestroika tuned into social aspirations and constantly taking up the thread of dialogue. As Manichean as this unconvincing claim was, it initially helped to bring calm to the country where elation was giving way to despondency. Nevertheless, the central press provided its readers with contradictory signs: according to some, everything would return to normal; according to others, Georgia would always be at the boiling point.[12]

Shevardnadze's mission bore some fruit. Provisionally, public order appeared to be reestablished; a new team at the head of the party and government was busy bandaging the wounds and placating the republic and its leaders. The time had come for the investigators to begin their work.

Questions Without Answers

It was not easy to decide who should be assigned to investigate the tragedy in Tbilisi. Few institutions or persons would be able to give an account of it without immediately being disbelieved by the people. Investigators of various origins crossed paths on the track of a truth that was nowhere to be found.

Just after the events, the Supreme Soviet of Georgia named an investigating committee to hear all concerned parties. This committee, headed by Tamaz Shogulidze, a well-known lawyer and specialist in human rights, heard blunt answers to some sensitive questions. Then the Congress of People's Deputies later assigned the liberal lawyer, Anatoly Sobchak—who in 1990 would be elected mayor of Leningrad—to conduct a second investigation.[13] Finally, an independent Georgian journalist formed his own committee, and its work contributed considerably to unearthing the truth. Thus, the experts tried to reconstruct the facts and answer three fundamental questions: Where and by whom had the decision been made to send troops against the demonstrators? Who had planned the military operation? Had the violence been intended or inadvertent?

To answer these questions, the committees heard the acting military commanders in the Caucasus, the republic's political leaders, experts in ballistics, chemists, physicians, and so forth. The committees proved unable to reach clear conclusions not because they failed to summon and hear innumerable witnesses but because of a negative climate from which glasnost had been consistently banished. Like most of the dramas that devastated the republics, from Chernobyl to the bloody clashes of the Transcaucasus, the Tbilisi affair was shrouded in mysteries, half-truths, and silence at every level. From Gorbachev to the army to the Georgian leaders, those responsible generally portrayed themselves as victims of insufficient information or simple excesses. It is also important to note the gap in information and glasnost between the accounts of the work of the investigating committee published in the Georgian press, in its Russian-language counterparts in the republic, and in the central press. For someone who wished to be genuinely informed, only two possible sources existed: the official Georgian daily, *Kommunisti,* and the *Moskovskie Novosti,* which tried to clarify the debate by sifting through the available information. Soviet liberal circles took a great interest in the Tbilisi affair as a test case for the future of perestroika. As support for this, Andrei Sakharov arrived in the Georgian capital to attend the hearings. But even his presence was not enough to create the climate of openness needed to answer the most serious questions.

At the heart of the debate was responsibility for the decision to send troops against the demonstrators. Was it a central directive or panic by local authorities? If the former, what factors would have led Moscow to make such a decision? Onto this initial question was grafted another: Who did it? If everything was decided in Moscow, who was to blame? Was Gorbachev responsible for the massacre, or was he one of its victims and the entire incident designed to discredit his policy? The stakes in this debate were clear. Gorbachev's liberal image could have been shattered by any conclusions directly implicating him in the tragedy. The size of these stakes probably explains the muddled conclusions of the Georgian committee.

The answer to the questions "who" and "where" followed almost automatically from the earliest explanation provided by Shevardnadze during his peace-seeking mission to Tbilisi. His thesis was upheld officially. Everything had been decided in Tbilisi, in a local politburo that panicked when it realized that it was losing control of the situation. On April 8, the politburo appealed to Moscow, asking for the intervention of the armed forces. According to Shevardnadze, the local politburo nearly unanimously supported Patiashvili's appeal for help. The chief exception, he said, was that of General Rodionov, a Russian and commander of the Transcaucasus forces.[14] A few days later, General Rodionov confirmed Shevardnadze's statement, arguing that it was not the army's role to maintain order and that the MVD had sufficient troops for this task.[15]

In an interview granted the *Ogoniok* in March 1990, Shevardnadze repeated his earlier statement, describing the conditions of Gorbachev's return home on the evening of April 7 from a trip to Great Britain, and explaining that when his colleagues who met him at the airport told him of the turmoil in Tbilisi, the secretary-general had insisted on the necessity of a political solution and then proposed to send him and Razumovsky to help the local government to resolve its problems. Tbilisi rejected this offer of mediation as inopportune, and it was not until the morning of April 9 that Gorbachev was informed of a military action of which he was entirely unaware. Shevardnadze concluded: "I can say in all good conscience that our head of state played no part in the decision to send troops against the demonstrators, not to mention the use of violence."[16] And he added that if the local leaders, who alone had made the decision, had then been unable to intervene between the crowd and the armed forces, it was because the situation had frightened them.

Not all of the various investigating committees found local authorities responsible for deciding to suppress the demonstration. In Georgia itself, Zhivi Gumbardidze, who succeeded Patiashvili as head of the party, cast doubt on the unanimity of the politburo (of which he was a member). According to him, this authority was not unanimous; the leaders had merely been informed that martial law would be proclaimed if necessary. On April 8, however, at the time of the meeting to which Gumbardidze referred, two envoys from Moscow arrived in Tbilisi: Kochetov, first vice minister of defense in the USSR, and Lobkoa, a representative of the Central Committee.[17] They may have come to Tbilisi in response to Patiashvili's appeal or to deliver instructions from Moscow. At any rate, the Shogulidze Committee confirmed Shevardnadze's claim on one point: they did not formally accuse Gorbachev but clearly stated that the operation had been decided on and controlled by Moscow. For its part, the unofficial parallel committee of the Georgian journalist concluded that responsibility was shared; it said that Moscow and the "highest officials of the USSR" were fully informed of the repression and the preparations for it. In fact, on April 7, in Gorbachev's absence, the Politburo of the Communist Party of the Soviet Union debated the situation in Tbilisi and possible responses to it. The committee's most serious accusation against the central government was directed at a close associate of Gorbachev, Lukyanov, vice president of the Congress of People's Deputies. In a telegram to the Central Committee of the CPSU, Patiashvili had mentioned that martial law was possible and that legal arrangements were being prepared—arrests, the control of the news—and his telegram in fact constituted a request for *implementing* the proposed measures. According to the investigating committee, Lukyanov read to the deputies from a slightly modified statement indicating that it was a matter of *informing* the Soviet Politburo of decisions already made in Tbilisi and not a request for authorization to make them. If this was the truth, Lukyanov, following the example of Pontius Pilate, knowingly created the ambiguity in order to relieve Moscow of responsibility for what was planned in Tbilisi. General Yazov, who was present at the Politburo meeting in Moscow at which the problem was debated on April 7, had that very day sent his vice minister to represent him in Tbilisi.

Under these conditions, it is not reasonable to believe that Moscow knew nothing about the repression that was being prepared in Tbilisi—details of which were given in the telegram addressed to the Politburo and read by Lukyanov—or about the meetings of the Georgian Defense Council and Politburo attended by General Rodionov, General Yazov's representative, at which the operation of April 9 was planned. At best, there may have been a desire to be unaware of it. According to Shevardnadze, the investigators never directly accused Gorbachev of taking part in the decision, but two men proved much more accusatory—Yegor Ligachev and Boris Yeltsin, who unreservedly linked Gorbachev to this decision.[18] Shogulidze and Sobchak were definite on one point: three important members of Gorbachev's entourage were directly implicated in the preparatory phase of the repression—Lukyanov, the minister of defense, General Yazov, and Viktor Chebrikov, who presided over the Politburo at the meeting of April 8. If this is true, it indicates that Gorbachev lacked authority over his close collaborators: he was informed only on the morning of April 9, on his return from London, of a violent repression that he had in principle ruled out, while on April 7, these three high officials had at the least known about the preparations for it.[19] More serious is the fact that Gorbachev never criticized or penalized any of them. Quite the contrary, a year later, General Yazov was promoted to marshal. This silence about their responsibility—at the very least in the withholding of information and the lack of any attempt to halt the repression that was underway by April 7—clearly indicates a desire by the center to avoid a full investigation. It seems difficult to dispute that the decision to intervene, even if it could be attributed to the local authorities, was approved in Moscow at a very high level by both civilian and military authorities. The parallel investigating committee in Georgia concluded in plain language that the decision had been made with the approval of Viktor Chebrikov, a member of the Politburo, and on the orders of General Yazov.[20] This accusation against the central government, which was printed in the Georgian press, was not directly disputed in Moscow, even if Shevardnadze, in an interview, insisted that the army, notably General Rodionov, and the local political leaders were responsible. As cautious as it was, Shevardnadze's position at the same time did justice to the Sobchak committee, which stressed the irresponsibility of the decision to intervene.

There was broad agreement on two points.[21] First, the Georgian situation did not call for a show of force. The crowd's totally nonviolent nature was shown by the things the demonstrators left behind at the time of the attack—notebooks and schoolbooks, purses and shoes for women and even children, and so on. But no weapons or other objects suggested violent intentions. On the other hand, the main question—who planned the repression?—still had no clear answer and raised an additional question: was the violence foreseen, and had it been provoked?

Concerning the plans for the operation of April 9, the information obtained by the various investigating committees pointed directly to two men, General Rodionov and General Yefimov, the chief of staff of the troops of the minister of the interior. But one question remained: had the generals been simply the executors or the real initiators of the plan implemented? General Rodionov flatly refused to appear before the investigating committees. On the other hand, he granted interviews in which he invariably stressed his own reluctance to engage in an operation of maintaining order and implicitly attributed the responsibility to General Yefimov.

The most troubling question is whether this violence was deliberate or accidental. On the evening following the violence, the Georgians made two accusations against the armed forces: the demonstrators had been provoked to panic, which triggered violent behavior in the troops; and the weapons used against them indicated the intent to injure and not to restore public order.

The first accusation was supported by the threatening behavior of the troops and the lack of a waiting period between the announcement of curfew and the movement of the troops. The Georgians were shown to be correct as were the Sobchak committee and the parallel Georgian committee, which the government eventually had to accept. Shevardnadze best summarized the first part of this argument:

> On April 8, the military organized a *rehearsal.* In the streets they deployed tanks and armored vehicles that frightened the population and made it fear that the small group of demonstrators who were permanently occupying the square would be attacked. That is why the whole city headed for the square that evening to see that its own were protected, convinced that a demonstration of nonviolent solidarity would prevent any repression.[22]

Thus by Shevardnadze's own confession, a needless deployment of force—the three committees admitted that the calm of the unarmed crowd did not justify this deployment—provoked a massive outpouring of people on the streets. And there, the curfew was decreed just minutes before it was enforced. Since a sizable crowd could not disperse in a few minutes, this short notice represented a scandalous provocation.

Some minutes after the curfew was announced, the troops went into action. Here again, two aspects of the action, at first rejected but duly established by the investigators, give the Tbilisi repression the dimensions of a premeditated massacre: they concerned weapons and toxic gasses. The demonstrators were attacked with finely honed sappers' shovels, as autopsies revealed. Moreover, for the first time, toxic gasses were used to disperse a nonviolent demonstration. Initially, the military and civilian authorities, like the first official information, denied using these unheard-of and illegal means to pacify a crowd. But proliferating eyewitness accounts and the investigators' observations eventually prevailed. When the accusations could no longer be denied, the question asked was, Who ordered the use of these weapons?

Once again, glasnost did not emerge from this affair unscathed. The Georgian political leaders, headed by Patiashvili, affirmed—and the investigators from all the committees took them at their word—that they had received assurances from General Rodionov and vice minister of defense Kochetov that the military action would have no victims and that the only equipment the troops would have at their disposal would be riot shields and police clubs. Someone decided to replace these means for safely dispersing demonstrators with deadly shovels, but no committee clearly established who. The Georgians, however, would not attribute the initial provocations or the substitution to local authorities; they felt that the central government was responsible for them, as well as for the use of toxic gasses. The Soviet army, through General Yazov, blamed the minister of the interior. The minister of defense told investigators that the army, unlike the MVD, had no chemical weapons of this type and that General Yefimov must also answer for the repressive means used.

When one draws up a balance sheet of truths and countertruths told during the investigation of the Tbilisi affair, two preliminary remarks are called for. First, obstacles were encountered by those who tried to understand the entire incident. Lead players in this tragedy refused to appear before the investigating committees and preferred to grant interviews in which no one could dispute their version. This was General Rodionov's approach. Conflicting claims were also presented without a comparison and analysis that would allow the truth to emerge. Military leaders accused the MVD of the worst excesses. Leaders of the MVD said that they knew how to intervene while ensuring people's safety, but they refused to debate the nature of the gasses and other repressive means employed. Shevardnadze put all his energies into clearing Gorbachev of any share of responsibility, and Gorbachev, for his part, was content simply to deplore the events.

The truth made but slight progress in this game in which all sides strenuously tried to impute responsibility for decisions and actions to a group other than their own. When it is reduced to the truth as told by the government, the Tbilisi affair represented a step backward for glasnost.

Second, for the Georgian people, whose reactions were practically unanimous, responsibility for the drama lay with Moscow, symbolized by the central role played by Russians—Rodionov, Yefimov, Yazov and his representative in Georgia, Kochetov, and Nikolsky, the second secretary (Russian, as he had to be) of the Georgian Communist party, whose authority in Moscow ranked him far above Patiashvili—in the preparations for "black Sunday." Because the obscurity surrounding all the official versions of the drama was never cleared up, and because the local political leadership was discredited, it was anti-Russian, not just anti-Soviet, feeling that won out over any other emotion after April 9. Before that date, the Georgians had little love for the Russians—old disputes went back to the incorporation of their country in the czarist empire in 1801—but they were accustomed to living side by side with them. After April 9, maintaining this common life became a serious problem; awareness of this hurdle was rapid and perceptible.[23] Only in Moscow was it for a time ignored.

The hypothesis that emerged from the most conflicting analyses, almost all doubtful in some details, was basically as follows:[24] The local leadership of the Georgian Communist party, worried by a movement whose extent and potential for growth were unknown to them, asked Moscow for political and military support for confronting possible excesses. The list of measures the Patiashvili team submitted to Moscow included martial law, the arrest of the ringleaders, and more generally the prohibition of all informal groups. But it probably was in Moscow that the decision was made to give a sharp response without waiting for the movement to develop. Everything prompted this choice. For geographical and human reasons, Tbilisi lay outside the view of international opinion. There was no large diaspora with permanent links to the republic to plead Georgia's case abroad. Georgia was growing increasingly restless, and demonstrations for independence, which were still extremely rare in the USSR of 1989, risked spreading to other republics. It seemed prudent to nip in the bud a possibly contagious example. Finally, the Caucasus was aflame. Although for various reasons the Armenians and Azeris had to be treated with care, they could not remain unaware that, despite efforts toward democratization, at any time a demonstration in a neighboring republic could be repressed in the USSR. Georgia could show that its devastating interethnic conflicts made it easy to justify the measures taken to protect the weakest ethnic groups.

If the repression had not been so savage, perhaps this approach might have succeeded.

Do the violence and the concealed provocation mean we should accept the claim of a "conservative" plot to destabilize Gorbachev? This theory would be plausible if it had weakened Gorbachev's share in the balance of power or caused a reaction by him against his supposed adversaries. But "black Sunday" produced no change in the balance of political power in Moscow. It simply led to a change of the team in place in Tbilisi, a team that had no influence on the leaders of the USSR. The only consequences of this affair were that in September 1989, General Rodionov left the command of the military district of the Transcaucasus, long enough after the events that it could not be inferred that this was a penalty; and on April 2, 1990, one year later, the Supreme Soviet voted for a law on the rights and duties of the troops of the MVD in the maintenance of public order.

Most of the provisions of this law implicitly pertained to the Georgian affair and attested to Moscow's desire to avoid a repetition of the excesses committed in Tbilisi. The law took away from the minister of the interior the freedom to dispose of troops of the MVD freely as he wished. A complex procedure for using troops was enacted: a republic's Council of

Ministers or other committee authorized by the Council of Ministers of the USSR would have to file a request with the central government, and the final decision would involve a presidential decree. The law also specified the kinds of weapons that could be used in operations to maintain order. The idea of forming special units for putting down uprisings at the periphery was at the heart of this law, and this idea sparked lively debates.[25]

The investigative efforts of the most liberal newspapers focused on the role of people like Chebrikov, the former president of the KGB, whose responsibility was the subject of a long investigation by the *Moskovskie Novosti.*[26] According to the newspaper, it was only after consulting with him that Patiashvili decided to ask for military reinforcements. Chebrikov was not accused of playing a major role in the events, however, and it is significant that Shevardnadze, clearly Gorbachev's spokesperson, remained silent on the matter.

What the tragic choices of 9 April showed, once again, was Moscow's incomprehension of the complexity of political developments and situations at the periphery. The choices indicated an uncertain policy that constantly vacillated between mollifying words and inappropriate actions. Above all, they were indicative of a flawed central decision-making apparatus that allowed some authorities—the army, the ministry of the interior, and so on—to take actions with grave consequences that compromised Gorbachev without later public condemnation.

More than anything else, the "black Sunday" of Tbilisi reveals the limits of Gorbachev's actual power. Although no one seemed able to infringe on his foreign policy turf, anybody and everybody judged themselves justified in encroaching on his authority in domestic problems. Gorbachev's acceptance of this, the unanswered questions, and the violent military actions against peaceful demonstrators indicates the limits of his power. "Black Sunday" was less an effort to destabilize Gorbachev's authority than an event that made it possible to measure, behind the myth, the reality of a power that only partially dominated a huge country and a complex bureaucracy.

The great loser in this crisis was the Soviet government, whose framework was the empire, and the notion of the Soviet people. After April 9, the Georgians, bad boys of the empire but still part of it by choice or by force, left it psychologically and wondered about how to leave it for good. And the sappers' shovels definitively destroyed the concept of the Soviet people. The conflict on the bloody square of

Tbilisi began with brothers on bad terms but soon became people who now were implacable enemies.

"A Good Georgian Is a Dead Georgian"

The Tbilisi military action was designed to calm Georgian national passions and also to demonstrate the central government's ability to attenuate interethnic conflicts. It achieved none of these goals, as subsequent events showed. Far from being subdued, Georgians concluded that only weapons could make it possible to avoid future violence. Acquiring new weapons became an obsession in every camp. Certainly, bearing arms and trafficking in them were not new in the USSR, particularly in the Caucasus, but at the end of 1989 there was a furious and nearly open race everywhere to acquire them.[27] Three months after "black Sunday," the Soviet ministry of the interior announced the seizure in Georgia of large quantities of explosives and firearms, including machine guns. Although the official communiqué said nothing about the provenance of these weapons, the press claimed that as many had come from outside Soviet territory as had come from thefts within it and that some had been bought from the military. A short time later, in Abkhazia, the MVD confiscated weapons and demanded that people relinquish to the authorities all those that remained concealed. The population's obsession with obtaining military equipment was met by a parallel government obsession with monitoring a society organized into networks for self-defense. The distance from repression to civil war seemed to be shrinking.

But the threat of civil war, or at least a scuffle, was also an internal problem in Georgia, one intensified by the growing agitation in the USSR in the late 1980s. Georgia, it must be remembered, is a composite state. Until Sunday, April 9, only the Abkhazians had made strong demands for the right to autonomy, but the events of April and the radicalization of Georgians soon led Ossetians to enter noisily on the scene.

In the summer and fall of 1989, the Abkhazian conflict returned to the forefront. Like the Georgians, the people of the autonomous republic were equipping their troops with weapons and ammunition; violent clashes proliferated and petitions were sent to Moscow to obtain the right to rejoin the Russian republic. The result of these clashes was dramatic: during a few weeks in July, there more than twenty deaths and more than 400 people wounded, and Abkhazians raided the premises of the MVD for the weapons stored there. A strike paralyzed public transportation in Abkhazia and thereby slowed down the entire Georgian economy.[28]

The persistent Abkhazian claims aroused other nations to express their will. Within its frontiers, Georgia includes a large and rapidly increasing Azeri minority—307,500 people in 1989—which was suddenly aroused from its age-old tranquillity by the rise of Muslim national feeling throughout the USSR. Accustomed to living among the Georgians, the Azeris suddenly discovered that they were different, and demanded a special governmental status, on the model of the Abkhazians.

It seemed entirely possible that Georgia would fall to pieces. The Abkhazians and Azeris were not alone in demanding the right to choose their future; in addition, they were supported by the Ossetians, who raised for the republic a problem that had been more or less forgotten in the course of time. The Ossetian question, which is to some extent reminiscent of that of Nagorno-Karabakh, is another inheritance from the complex federal structure elaborated by Lenin and Stalin. The Ossetians are not a negligible group. Six hundred thousand of them lived in the USSR, divided between two states: Russia, where there are 335,000 of them in the North Ossetian Autonomous Republic, and Georgia, where they number 164,000, of whom 65,000 live in the South Ossetian Autonomous Region,[29] with the rest scattered across the territory of Georgia. The fate of this people, which was united to Russia by Catherine II and has shown itself over the centuries extraordinarily loyal to the empire, has been in the most striking contradiction to its merits. In the past it provided heroic officers who distinguished themselves on every front where the rulers of Russia made war. The Soviet government honored their ferocious resistance to the German occupiers, rewarding them in 1944 with a substantial enlargement of the territory of the autonomous republic forming part of Russia. Why, then, was this people cut in two, divided between two Soviet republics of unequal status: an autonomous republic in Russia, and an autonomous region in Georgia?

Despite a common origin, the Ossetians (an Indo-European people speaking a language of Persian origin) are divided religiously, mostly Muslim in northern Ossetia and mostly Christian in Georgia. Both groups, however, are solidly attached to Russia. In all, 70 percent of them speak fluent Russian, while this is true of only 33 percent of the Georgians. Even those living in Georgia have mastered Russian in slightly greater numbers than the Georgians.[30] Their aim has always been to be united in a single autonomous republic; they consider Russia, their accepted protector since Catherine II, the appropriate framework for such a union. As early as 1925, the leaders of the two Ossetian communities had tried to convince the Soviet government to make a decision favorable to union with Russia, but Stalin would not budge. Georgian in origin and Ossetian through his mother, he refused to spark an explosion of fury among his Georgian compatriots, mortally wounded by the forcible termination of their independence in 1921, and whose revolt in 1924 ended in a bloody repression. The Ossetians were not reunited, but the Georgians did not forget that, during a terrible repression, the Ossetian people, far from supporting them, turned against them.

Relations had never been easy between the autonomous region and Georgia. The Georgians felt that the Ossetians were unsure of themselves and readily manipulable by Russia and that therefore the gradual destruction of their identity was necessary—particularly the ties binding them to the Ossetians in the north that made them want to be reunited. The only arrangement that Georgia could accept would be the incorporation of a reunified Ossetia within the Georgian state. For their part, the Ossetians showed persistent hostility, fiercely defending every component of their identity—language, traditions—and maintaining that "the only good Georgian is a dead one." This was also the creed of the Abkhazians. Within their respective states, both groups endeavored to make life impossible for Georgians living there.

For decades, Georgia was able to keep these turbulent citizens quiet. But the crisis of 1989 suddenly made all the conflicts resurface: April 9 made it possible to measure the Ossetians' hostility to the state in which they were incorporated. Far from sharing the same sense of oppression and identifying themselves with the tragedy in Tbilisi, Ossetians exhibited a splendid indifference and even a certain satisfaction. In all the clashes, their solidarity was with Georgia's opponents—both the Abkhazians and the Russians. In the months following the Tbilisi massacre, the press stressed how much these demonstrations of "desolidarity" (not just indifference or passivity, but active hostility) had offended the Georgian population.

From then on, the positions of both groups hardened. During the summer of 1989, a Popular Front was born in the Ossetian region, and it began a campaign—petitions to the authorities of the USSR, Russia, and Georgia—for secession from Georgia and the unity of the Ossetian nation within Russia. The possibility of a renewal of Soviet federalism and the acute sense—first arising from the Armenia-Azerbaijan crisis—that the map of the USSR was to be redrawn impelled all the opponents of the existing system to present their demands. And the separated Ossetians had rights to assert. By the end of the 1980s, however, the Soviet mood was not propitious for level-headed discussion. Every demand was couched in violent terms, and its proponents, at least in this part of the USSR, were convinced that obtaining a hearing required being armed and arousing fear. Like the Georgians and the Abkhazians, the Ossetians armed themselves, and the general insecurity—for they obtained arms mainly by violence and theft—increased. When people are armed, they are tempted to use their weapons—and the theory of the "good dead Georgian" was put into practice.

It became dangerous to be a Georgian in an Ossetian setting, so much so that in the fall of 1989, the Georgian minister of the interior had to dispatch troops to Ossetia. The Ossetian Popular Front, aware that Georgia would not accept any loss of territory, then issued another demand: if their borders were not redrawn, the Ossetians in Georgia were to be given the same status as those in Russia; Ossetia would switch from an autonomous region to an autonomous republic, which would increase its rights in every sphere—political, cultural, and economic.

That was exactly the opposite of Georgia's intent. Obsessed with the threat of territorial dismantling that the minorities held over them, the Georgians were committed to a policy that consolidated their own preeminence, notably by promoting Georgian as the state language and hence the language of integration and by affirming the total supremacy of Georgian law. The language battle was not just symbolic but reflected the balance of power. Ossetians demanded that their own language have equality with Russian and Georgian in the region and refused to accept that throughout the republic all official documents be printed and business be transacted in Georgian. By insisting on trilingualism, they gave the Georgian population a group status equal to the others in the republic. From there, it would be an easy step to affirm that Georgia was merely a conglomerate of ethnic groups, as in the Soviet federation, where no group was above the others. This was understandably worrisome for Georgia, particularly when the rapid breakup of the USSR suggested a parallel breakup of all the governing structures in which numerous nationalities coexisted.

The government in Tbilisi continued to resist the mounting Ossetian demands as well as those of the Abkhazians and the Azeris. It rejected all demands for a change in status, imposed rulings that allowed the

Georgian language to gain ground, and, above all, committed itself to a constitutional revolution to protect itself from pressures from Moscow favoring the minorities. On November 18, 1989, the Supreme Soviet of Georgia voted for constitutional amendments giving it the right to reject any federal law that was contrary to the republic's interests.[31] Their intent could not have been plainer—to forestall any changes that Gorbachev might make in Soviet federalism.

But in rejecting the primacy of federal law, Georgia was also rejecting federalism and hence the federation. In April 1989, through an action as savage as it was unjustifiable, the central government had hoped to cool Georgian nationalism. Six months later, in the face of the onslaught of small nations demanding the right to emancipation, Georgia concluded that to preserve its integrity, it first would have to distance itself from the USSR. The "lesson" that Moscow had wanted to give it on April 9, which had no doubt helped to exacerbate internal rifts, whirled back like a boomerang against its originators. For Georgia, everything then came down to what in Tbilisi was called the *Russian question.* Injured by a power it identified with Russia, threatened with loss of territory for the benefit of Russia and by what it took for Russian manipulation, Georgia henceforth applied to Moscow the idea that its minorities applied to it: the only good Russian is a dead Russian.

6

The Suitcase or the Casket?

THE riots in Alma-Ata were clearly colonial riots pitting the peripheral regions against the central government and the Kazakhs against the Russians. In the Caucasus, nations clashed and continued doing so over their borders and over the control of territories and populations. These conflicts had originated in the irredentism of peoples who rejected the territorial status imposed on them by the nascent Soviet government in the early 1920s. There remained a final category of interethnic conflicts that did not challenge the central government—the hatreds created by population dislocations. The violent incidents unleashed in Central Asia starting in 1989 were all caused by the rejection of immigrants, which led to their cold-blooded massacre.

Hatred of the Other

The succession of Central Asian crises displayed the same logic—the crystallization of national feelings as well as economic and social anxieties about immigrants. The best-known example of this type of conflict, the one that brought it to light, was the June 1989 massacre of the Meskhes living in the Ferghana Valley. Actually, however, this massacre was merely one link in a chain of similar violent episodes.

In June 1986, at Dushanbe, the capital of Tadzhikistan, several thousand Tadzhiks attacked all those who looked to them like foreigners. The USSR, engrossed by the changes going on in Moscow and still unaccustomed to the free flow of information, paid no attention to these incidents in which many people were wounded.

In October 1986 the action switched to Frunze, the capital of Kirghizia, where Kirghiz students attacked non-Kirghiz students. Violent incidents recurred some months later.

The Alma-Ata explosion was followed by two years of calm during which the inhabitants of Central Asia, accused of corruption and other immoral behavior, seemed to endure passively the discredit heaped on them. They were the "villians" of a USSR that dreamt of purity, and their silence sharply contrasted with the mounting uproar in the Caucasus. In December 1988, however, this region too was the scene of disorder. Uzbekistan—a scorned republic in which a real climate of terror reigned—was the main region affected, even if it was not the only one.

In December 1988 and February and April 1989, pitched battles bloodied Tashkent. Students and workers clashed at the university and in the factories and streets, armed with iron bars, nightsticks, and sometimes light weapons. The Uzbeks also went after the same victims—the *foreigners,* whoever they might be. The same hostility toward foreigners sparked the violent incidents that were repeated in February 1989 in Dushanbe, where school pupils and university students assaulted anyone who did not understand their language.

Up to this point, the events did not go beyond a certain xenophobia, on which the USSR has no monopoly. But with the larger clashes in Turkmenistan, Uzbekistan, and, again, Kazakhstan, the nature and magnitude of this phenomenon were becoming clearly worrisome.

In May 1989, in Askhabad, the capital of Turkmenistan, violent demonstrations broke out in the oil-drilling center of Nebit Dag. Apparently, the movement was spreading to other parts of the Soviet Union: popular hostility was vented on cooperatives and small entrepreneurs whom a destitute society took for revivals of a system in which neighbor was exploited by neighbor. But in these two cities, the violent incidents had something in common: the terrorized members of the devastated cooperatives came from the Caucasus; the slogans, "Down with tight-fisted employers!" "Close the cooperatives!" "Get the speculators!" were everywhere punctuated by a revealing shout:[1] "No Caucasians or Armenians in Turkmenistan!" The instigators of the violence were primarily young people, which made it possible to blame the affair on hooliganism, a heritage of the moral laxity of the years of stagnation. But despite this classic explanation, frightened Turkmenistani authorities called a meeting of the representatives of all the ethnic communities living within their territory. They intended to affirm interethnic solidarities and overcome the divisions in the hitherto peaceful republic. This desire for peace explains why, just after barely being noticed, the violence in Ashkabad and Nebit Dag was immediately forgotten. Also contributing to the forgetfulness was that the fairly heavy toll in physical damage did not include many victims.

Quite different was the toll of the violence that bloodied the Ferghana Valley three weeks later. The incidents broke out on May 23 in the little town of Kusavia some nine miles from Tashkent. The reason for it was trivial. A young man protested that the strawberries for sale at the market cost too much, and he overturned a basket of them and manhandled the woman selling them. A veritable battle started up around the market, and one person died. The next day, the battle resumed: a second death and more than twenty severely wounded were added to the first day's toll. Despite efforts by the police to prevent it, the agitation spread to other cities in the Ferghana.

On June 4, the clashes turned into riots. Gangs of young Uzbeks armed with chains, axes, iron bars, and clubs plundered and set fire to the houses of the Caucasians, striking, mutilating, and destroying everything in their way. They attacked public buildings, chiefly police stations, where they tried to seize weapons. At the Party headquarters in the city of Ferghana, they took the two first secretaries as hostages. The police were powerless in the face of a blind fury that was spreading throughout the whole region, even though they and the troops of the MVD had been reinforced by regular army units. Despite this deployment of the armed forces, the riot continued to spread over an ever larger area and reached the city of Kokand. Automatic weapons appeared. A train was stopped, and its load of gasoline poured onto the track. The rioters threatened to blow up the train and kill their hostages if the authorities did not accede to their demands. The entire republic was ablaze, public transportation was immobilized, and the number of victims was growing.[2]

The USSR was perhaps beginning to get used to violence at its borders, but until June 1989 the phenomenon might have seemed limited: an upheaval in Alma-Ata, demonstrations put down and thereby temporarily overcome in Georgia. Only the Lebanonized Transcaucasus could call up the case of Uzbekistan—racked by uprisings, ignoring the police and military, defying the government, and imposing the will of the mob.

In Uzbekistan, popular fury turned against the Meskhes, an immigrant community of people from the Caucasus once deported by Stalin and now forbidden to return to their homeland. The Uzbeks suddenly began hating these immigrants so intensely that the incident of May 23—the young man who found the strawberries too expensive—degenerated into a systematic massacre. Not content with burning their houses, violently assaulting them, and attacking camps in which the authorities had hastily settled them for their protection,[3] Uzbeks demanded the immediate departure of all Maskhes from the republic.

Unlike the Turkmenistan riot, the toll was heavy: a hundred dead, more than a thousand wounded, hundreds of houses and public buildings burned, immeasurable destruction, and, above all, 34,000 Meskhes forced to flee to other republics. The return of calm was only momentary. The Uzbeks continued to harass the Meskhes who had stayed in Uzbekistan protected by the police; periodically, groups of several hundred demonstrators launched assault against these unfortunate people. The persistence of crises and violence confirmed that Uzbeks would not stop until the last Meskhes had left their soil. Even in Samarkand, which had hitherto remained calm, the population attacked the Armenians at the beginning of 1990. But Uzbekistan had become so accustomed to ethnic violence that when it reached this city it went on practically unnoticed.

Barely had the fire been put out in Uzbekistan than another broke out in Kazakhstan, in the city of Novy Uzen, an oil-drilling center. The violence here was less severe than the attack on the Meskhes—ten dead, between 100 and 200 wounded, physical damage—but its causes and effects were the same as in Uzbekistan. Here, the targets of the local population's fury were immigrants from Daghestan. The scenario of the crisis was nearly indistinguishable from the one in the Ferghana. It all started with a minor clash between young Kazakhs and Caucasians at a public dance on June 16.[4] The incident quickly turned into a confrontation between ethnic groups. Young Kazakhs overran the streets of Novy Uzen, setting shops on fire, clubbing everyone who looked Caucasian, attacking public buildings, and storming police stations for coveted weapons. For several days, railroad travel was paralyzed, businesses were closed out of solidarity or caution, and public demonstrations were held and attended by women and children. Overwhelmed despite the reinforcements, the police and the military shot into the air, wielded clubs, and used tear gas but failed to check the movement, which was spreading to the far western part of the republic on the border of the Caspian Sea.[5]

The Kazakhs were here willing to employ a new strategy in fighting the police. They were grounded by the curfew imposed on Novy Uzen on June 19, but demonstrators circumvented the problem by moving the conflict from the city where it arose, which was now under heavy surveillance, to towns leading to the Caspian. They exported violence, economic paralysis, and terror directed at immigrants from the Caucasus. Not content with attacking the oil fields of western Kazakhstan, they also expanded the strike to the eastern part of the republic, as far as Alma-Ata, the starting point for all the ethnic revolts three years earlier.

Kazakhstan, whose size seemed to condemn it to evolve in many different ways, was for once united in a hatred of immigrants through the systematic activity of the rioters of Novy Uzen. Although this revolt cost fewer lives, it had substantial political effects on the region's unification as well as economic repercussions. First to pay the costs of the events was the oil industry. Production sank because of sabotage and the shutdown of the oil wells and the natural gas facilities.[6] In addition, the flight of competent technicians from the Caucasus was difficult to offset with unskilled Kazakhs.

Perhaps because of their scattered nature and the lesser number of victims, these incidents were probably of less concern to the central government than the nearly universal uprising in Uzbekistan. On June 12, Prime Minister Nikolai Ryzhkov and Viktor Chebrikov were in Tashkent—a sign of Moscow's anxiety. In Kazakhstan, the local leaders were content to name investigating committees.

But violence spread, sparing no republic. Long immune, Kirghizia too was infected with a hatred of immigrants. This hatred was demonstrated in January 1990, not against actual immigrants, but against rumors that some refugees from the civil war in Azerbaijan would be given asylum in Kirghizia. This idea incensed the Kirghiz, who occupied the vacant lots in Frunze where they suspected that the refugees would be housed, hastily built sheds, organized to defend themselves, and demonstrated en masse against any population movement into their republic. The people of Frunze were unconcerned that the potential refugees were Russians who had been made utterly destitute by the war in Azerbaijan but of whom no trace had yet been seen. The intensity of the response in Kirghizia showed that here too serious clashes could be expected. This was proven a few months later, in June 1990, when interethnic clashes occurred between Kirghiz and Uzbeks living in the region of Osh.[7]

Nor was Tadzhikistan spared interethnic violence. The war in the Transcaucasus cast its shadow over all the republics. On February 11, 1990, mayhem broke out once again in Dushanbe, where the Tadzhiks demanded the immediate departure of refugees from Armenia.[8] At first, the Tadzhiks used harassment, but the violence that left a few dead and many wounded soon turned to a state of war. While in Moscow Gorbachev was thundering against national passions and demanding that every means be used to restore order in a blazing Central Asia, neither sending in troops nor immediately imposing a curfew calmed things down. The attack on Armenians was followed by a true insurrection against the Soviet government. The Tadzhiks attacked the forces of the MVD and the army.

These clashes indicated how extensive revolts at the periphery had become since the events in Georgia. The time for iron bars was past, and insurgents were increasingly well armed. In the face of them, the regular army was reluctant to intervene. One unit sent to Dushanbe even refused, alleging that the maintenance of order was not its job. After the reservists refused to fight in Azerbaijan, the unwillingness of the regular units betrayed the disastrous effects of ethnic violence on Soviet institutions as well as on individual morale. When the police finally regained control of the situation, the toll was high: a minimum of twenty dead and nearly 600 wounded. The list of victims in the ethnic conflicts continued to lengthen.

At the Heart of the Conflict: Rejection of the Immigrants

For the Soviet government, the successive uprisings that had in a few months destabilized Central Asia officially were due to three causes. The first explanation was drugs, alcohol, and habitual criminality.[9] After the pogrom in Sumgait, all official news reports on ethnic violence mentioned these factors to distinguish demonstrators from the rest of the society. The second explanation was that these criminal types—sodden with alcohol and drugs—who started the violence had been manipulated; the violence served a plan and so represented a conspiracy[10] against perestroika and the Soviet government. By whom? Here, many possibilities were raised. The clandestine conspirators had used numerous and various means. The demonstrators were paid to demonstrate. The weapons, used everywhere, often came from the outside world. On this point the accusation was occasionally specific.[11] In Tbilisi, weapons were provided by neighboring Turkey and in Azerbaijan, by Iran, which denied the accusation. In Tadzhikistan, the military equipment was supplied by Afghanistan. The United States itself was at times accused of trying to destabilize the border regions to aid the Afghan resistance.[12] A third factor was cautiously advanced—Islam, particularly the extremist sects. But none of these explanations was persuasive, nor did they mention—and the government knew this—the two major problems of Central Asia: immigrants and the socioeconomic situation.

The complex human map of Central Asia had been considerably modified, especially since the war years, by three elements, all involving population movements—a massive and continual dispatching of Russians to the periphery; Stalin's deportation of entire ethnic populations, from the Germans of the Volga to the peoples "punished" for collaboration; and the final wave of refugees from the war in the Transcaucasus, for whom the Soviet government tried to find sanctuary.

The Russians who long were on the increase in Central Asia have for several years been heading home as demographic pressure and a growing nationalism combined to make their lives at the periphery harder and harder. We can see this at a glance in Table 6.1.[13]

Russians declined in absolute numbers in three republics out of five and proportionally lost ground in all five republics as the dominant local nations experienced a spectacular growth rate. Thus, in Uzbekistan in 1979, there was one Russian for every six Uzbeks; ten years later, one for almost nine. It was Uzbekistan that officially demanded a halt to the influx. Slogans such as "An end to Russian immigration!" and "Russians

TABLE 6.1
Number of Russians in the Republics of Central Asia

	Total Population in 1989	1979	1989
Uzbekistan	19,808,077	1,665,658	1,652,179
Kazakhstan	16,436,115	5,991,000	6,226,000
Kirghizia	4,257,700	912,000	916,000
Tadzhikistan	5,089,000	395,000	387,000
Turkmenistan	3,512,190	349,000	334,000

go home!"—so dear to the Uzbeks—also rang out in demonstrations in Ashkhabad, Frunze, and Dushanbe. But during the most recent violence, outcry against immigrants was directed primarily at deportees or people forced to flee wars raging in their own republics. Significantly, most of the deportees were Muslims, and the rhetoric of Islamic solidarity prevented them from foreseeing the massacres they would suffer. Conversely, although the Russians were unpopular, sometimes harassed, and always humiliated, their systematic massacre was not part of the picture, for the time being at least.

Although Germans had lived along the Volga for three centuries, Stalin feared that they formed a base for collaboration in the coming war with Germany and relocated them in Kazakhstan. A minority in Central Asia, they were a solid group of 956,000 persons in Kazakhstan in 1989 and attracted little or no hostility.[14] In no demonstrations were Germans jeered or their departure demanded. As many as could, however, left on their own: 5,000 have left Tadzhikistan in recent years, several hundred from Turkmenistan. They returned to the homeland that their ancestors had left during the reign of Catherine II, invoking their right to the "reunion of families" in the name of principles accepted by the USSR in Helsinki.

The uprising in Ferghana confirms that hatred can exist between Muslim peoples—that of Uzbeks for Meskhes. Meskhes are Georgian in origin, Muslim (the majority Sunni), and partly Turkicized. They had been snatched from their villages in central Georgia in November 1944 at Stalin's orders and herded onto the roads to Kazakhstan. Accused of collaborating with the Germans, they were among the "punished" groups denied the right to exist as a specific community. Khrushchev rehabilitated some of these punished groups and authorized their return home, but the Meskhes, like the Tatars and the Germans, were denied this improvement in status. Since 1956, following the example of the Tatars, they have fought to regain their land of origin. To be heard in Moscow, they lived in the hastily built refugee camps designed to save the lives of escapees from the massacres. Nikolai Ryzhkov, prime minister of the USSR, visited these camps, and for the first time since their deportation, Meskhes heard of a possible return.

Uzbekistan had a population of some 60,000 Meskhes[15] who worked, owned businesses, and held jobs coveted by the Uzbeks. Although Uzbek hostility toward Meskhes was founded in economics, it was exacerbated by their fierce desire to leave a Muslim country. The Tatars' determination to return to Crimea has in the past provoked the same reflexive outrage in Uzbekistan. Returning the Meskhes to their original homeland was difficult, however. Although the Uzbeks now wanted to be rid of the Meskhes, Georgians, who were grappling with their own minorities, let Moscow know that they were not disposed to see the number of minorities increase and that Georgia did not have the resources to receive them.[16] Central Asia has more than 150,000 Meskhes (more than 300,000, according to Georgian estimates; 143,862 inhabitants out of 160,000 in this region in 1989), who have their own language and Shiite religion under the spiritual authority of the Aga Khan.[17] In 1959, the Tadzhik authorities, who deny this people's national existence, officially declared the Meskhes to be Tadzhiks, thus making the Meskhes legally nonexistent in the censuses.[18] Although the peoples of the Pamir appeared resigned to this attempt at forced assimilation, they eventually were influenced by the heightened mood of nationalism in Gorbachev's USSR. The Tadzhiks brutally planned—logically enough, from their viewpoint—to wipe the autonomous region off the map. The seemingly inevitable territorial redistricting in the USSR led the people of Pamir to mobilize, making inevitable an explosion similar to the one in Karabakh.

Another mounting crisis stemmed from the hostility of the Tadzhiks toward the Uzbeks and neighboring Uzbekistan, which added typically Caucasian territorial and interstate conflicts to internal ethnic conflicts. The Tadzhiks, as a result of the rise in nationalistic tensions over the past few years, showed their hostility to the Uzbeks in three ways. First, within their state, Tadzhiks tended more and more to treat the Uzbeks living among them as intruders and as members of an inferior civilization. (The Uzbek population reached 1,197,000 in 1989, representing over a fifth of the region's population of 5,089,592, of whom 3,168,193 were Tadzhiks.) Tadzhiks claimed descent from classical Persian culture and considered their neighbors barbarians who must either accept the Tadzhik culture and language or go away. This dispersed community dreamed of returning to Georgia, where no one wanted them, and could not be satisfied by some temporary shelter in the Russian republic.

Victims of violence in Kazakhstan, the Caucasians—Ingushes, Balkares, and Karatchais—also represent a forced immigration dating from the World War II years. "Rehabilitated" by Khrushchev, most returned to the Caucasus; the violence committed against Caucasians who chose to remain reflected the Kazakhs' refusal to tolerate foreigners on their soil.

As recent immigrants fleeing the massacre in the Transcaucasus, the Armenians encountered racial hostility throughout Central Asia. Whether already present or rumored to be coming, they were systematically rejected in Frunze, Turkmenistan, and Tadzhikistan. Their misfortune met with no compassion, and the proverbially hospitable people of Central Asia uniformly rejected them. The slogan was universal—"No new immigrants."

In Tadzhikistan, the interethnic conflict was complicated by several problems. The demonstrations in Dushanbe in 1990 were dominated by the ubiquitous slogan that each republic adapted: "Tadzhikistan for the Tadzhiks!" This implicitly referred to the expulsion of the Russians—although they had begun leaving the republic on their own—and the banning of any newcomers. Added to these already well-known hostilities was the Tadzhiks' brutally assimilationist attitude toward the smaller groups in the Pamir who were in the majority in the autonomous region of Badakhshadn.

Second, the Tadzhiks were outraged at the fate of their conationals in Uzbekistan. There, in the face of 14 million Uzbeks, the 931,547 Tadzhiks complained of discrimination and worried that they would be assimilated into the Uzbek group, whose demographic dynamism fueled talk that it was becoming the region's melting pot. "Uzbek imperialism"—which the Tadzhiks claimed they were denouncing—was reflected in the classification of many Tadzhiks as Uzbeks on the basis of imprecise identity papers. The Tadzhiks thus accused Uzbeks of underhandedly using shady administrative procedures and of intensifying cultural pressures to assimilate minorities.

The third element of the conflict was the Tadzhik demand for the return of Bukhara and Samarkand, where the Tadzhik population and culture are dominant, to their republic.[19]

Central Asia was beginning to look like the Caucasus, and Moscow saw even fewer reasons to yield to the Armenians' demands concerning Karabakh as similar demands surfaced on the Asian periphery.[20] The spiral of violence in this region was particularly worrisome because it occurred later than elsewhere. Events in Frunze and Dushanbe suggested that an unverified or even inaccurate rumor—"They're going to ship the Armenians here"—could produce a blowup. Provocations—and in both cases, there was an obvious intention of stirring up the Kirghizians and Tadzhiks—could lead to massacres. It was also worrisome because, as in the Caucasus, the large nations now denied minorities the right to exist on their soil and challenged the territorial organization imposed by Stalin in the early 1920s. The elements of the tragedy were coming into place—pogroms of immigrants and wars between states over ethnic hostilities and rivalries. The sheer size of Central Asia, the large numbers of peoples involved, the prevalent social conditions there, and Islam made this region even more dangerous for the USSR than the Caucasus was.[21]

A World of Despair

Interethnic conflicts, especially when they break out suddenly, are primarily directed at the weakest group, the immigrants, and have social and economic roots. Hostility toward the Russians is a different kind of problem because it concerns people who are officially equal but who feel that one group is "more equal than the rest." This is particularly true of Central Asia, which in the mid-1980s emerged from a long period of quiet to a sudden awareness of all its problems. Everything led to the same observation—an accelerating decline in development. For a long time, the inflated assessments of local leaders gave the peoples of the region a sense of continual progress. Transfers of resources helped the region strengthen its balance sheets, which then misrepresented reality. Glasnost laid reality bare, but the disparity between the facts and the official whitewash was so great that truth seemed completely unbearable. Treated as pariahs by society, subjected to unending purges, accused of sponging off others, warned that they would have to survive on their own, confronted with economic collapse, the people of Central Asia were reduced to despair. An economic disaster loomed just as a demographic explosion intensified their needs and increased the number of young people for whom the market had no jobs.

The Soviet government had long urged Central Asia's labor force to emigrate to Russia or Siberia to find work. But the economic reforms of Gorbachev and his team undermined labor emigration: there was no place in the USSR for a superfluous labor force. The increasing populations of Central Asia were dogged by unemployment but were unneeded outside of their own regions. Employment was a major struggle in this region, where the demographic explosion had in less than two decades outpaced all predictions—in hiring, housing, and supplies. No preparations had been made for the near-doubling of the population (in Uzbekistan, there were 11,800,000 inhabitants in 1970 and 19,808,000 in 1989).

In fact, the single-crop cotton growing imposed by Stalin on a large part of this region had two effects.[22] Because it was largely mechanized, this kind of agriculture offered the population few jobs, and because it was carried out at the expense of food crops, this specialization made most of the republic highly dependent on the rest of the Soviet Union. In the late 1980s, when everything was disrupted in the USSR—such as transportation and warehousing—and food became scarce, specialization doomed Central Asia to near-famine. Moreover, the ecological disaster of the Aral Sea, which deprived most of the neighboring republics of drinking water, and the intensive use of pesticides created appalling prospects for the region's inhabitants.[23] This was certainly nothing new, and glasnost disclosed rather than caused these conditions. Now, however, because the need to keep silent has given way to the right and even the duty to speak up, everyone knows that water pollution and air pollution are higher there than the federal average and the area heads the list of regions threatened by these new dangers.

One reason that the nations of Central Asia are hostile to the settling or transfer of immigrants to their land is that the immigrants will obtain the best jobs because of the region's cultural backwardness. The western Kazakhs justified their massacre of the Caucasians by observing that foreigners got the well-paying jobs in the oil fields and the factories and created pressure to exclude Kazakhs. Caucasians deported in 1944, Armenians, and Azeris were accused of monopolizing the cooperatives where they set high prices. "We are condemned to a cotton industry that pays practically nothing," complained the Uzbeks at demonstrations, shouting, "Raise the price of cotton!" and "Drive out the thieves!" They added that in Ferghana, young Uzbeks were reduced to unemployment while immigrants—Meskhes, Tatars, and so on—sneered at Uzbek poverty. They also complained that while Central Asia lacked everything, the current disarray in the USSR led "foreign speculators" to set any prices they wished for food and most basic articles, thus placing Asians at their mercy. In the explosions of popular fury, the immigrants' shops, houses, and land were attacked, set on fire, and destroyed. The expulsion of those who lived there was a recurrent leitmotif throughout Central Asia.

Even though the people of the region, like the Tadzhiks and the Uzbeks, also fought with one another, their first priority remained to drive out all those who did not belong to the group. A poll taken in Kirghizia as to the state of interethnic relations in the republic provided an arresting view of the prevailing state of mind and can be extrapolated to all of Central Asia.[24] A majority of those polled condemned a policy of equality between the Kirghizians and minorities living in the republic, for "equality benefits minorities and disadvantages the national majority." The conclusion was that the minorities in the republic should not have the same rights as their protectors. They had to choose between assimilation—which they universally rejected and which was contrary to the Soviet ideology of the harmonious development of all groups—and the radical alternative currently going on elsewhere—"the suitcase or the casket."

Fraternal Islam?

"Every Muslim is the friend and brother of any other Muslim, whatever his nationality or land of origin. A Muslim ought never to suffer at the hands of another Muslim."[25] This precept, repeated many times in Soviet mosques, appears to have been wholly forgotten in the interethnic conflicts of Central Asia. Nevertheless, in many demonstrations in the region, the green flags of Islam alternated with placards bearing slogans, even when the demonstration was a proclamation of hostility to other Muslims. This situation was only seemingly paradoxical.

Soviet authorities worried that behind the violence, Muslim agitation, even Islamic propaganda coming from the outside world, profited from any discontent by destabilizing the periphery; they were partly, but only partly, right. Bonds were indeed growing stronger between Soviet Muslims and their neighboring fellow Muslims—Iranians and Afghans. Saudi Arabia was clearly taking an increasing interest in this part of the USSR, populated as it was by more than 50 million Muslims and contiguous with the center of the Islamic world. But nothing indicated that the mounting passions were being stirred up by outside sources. Propaganda tracts and tape cassettes from Iran and Afghanistan were circulated and were invoked to support the claim of foreign provocation, but they appeared primarily in the areas adjacent to these countries and not as frequently in the urban centers where the most serious disturbances occurred.

Nevertheless, it is impossible to contest Islam's recent growth in Central Asia. In Asia as in the Caucasus, Gorbachev's relatively liberal policy on religion resulted in the opening of new mosques and an increased effort to train more (and more effective) clergy.[26]

It will take time to change a situation that is still not prepared for

Islam's ascendancy over society. In 1986, the USSR had only 1,330 open mosques and nearly 2,000 mosques with no official status but functioning in a tolerated illegality;[27] some 20,800 prerevolutionary mosques served a Muslim population only a third as large. The two Islamic academies now operating in Central Asia—a third is planned for Azerbaijan—have inadequate resources for accommodations and education in view of the needs.

The winds of reform, however, have changed relations between Muslims and their hierarchy. Indicative of an authentic psychological revolution is the February 3, 1988, demonstration that gathered in Lenin Square in Tashkent, where more than a thousand people demanded the dismissal of the grand mufti Shamsutdinkhan Babakhan, who was accused of corruption and loose morals. The government had already accepted religion as a stabilizing influence on society, but it was doubtful that an Islam so contested by the people could play much of a stabilizing role, especially in such an unsettled moral context. Central Asian society must face not only a dramatic economic situation, involving an expanding population and deteriorating material and sanitary conditions, but also rampant moral decay. The young people are threatened by alcoholism, drugs, and delinquency.[28] The emancipation of women, once held to be one of the most glorious achievements of the Soviet system, is rapidly losing ground, and, as a corollary, women's suicide rates are climbing rapidly.[29] In a few years, social progress sponsored by the system (education, equality of the sexes, and the protection of children) and by conventional values (family cohesion, parental authority, sobriety, and so forth) collapsed, exposing, as elsewhere in the USSR, a society that has lost its traditional reference points and that, through the revelation of the Soviet lie, is persuaded that there are no limits to the satisfaction of individual desires. Appalled by the magnitude and consequences of this social disarray, Muslim political, intellectual, and religious leaders have turned to the Islamic value system as a basis for society's moral recovery and solidarity. Islam is to be a social glue for uniting the nations of this region, not an authoritarian or fanatical version of the religion. In this period of uncertainty, Islam is expected to play a role similar to that played by Catholicism in modern-day Poland.

Islam thus has become a mode of legitimizing the national state that wishes to survive and develop independently of a system that had sacrificed them to the Soviet idea of progress.[30] Thus understood, Islam unites only the members of the same national community and must be identified with their interests. As the profile of the nation regroups, more and more demonstrators are expected to rally under the green flag of Islam.

When despair causes a value system to be seen as a means of uniting a community against the outside world, notions of solidarity and peace mean little. The system then helps reinforce the individuality of each group. In this case, Islam provided the Central Asian powder keg with an exceptional potential for popular mobilization. By turning toward Islam, this region distanced itself from the Soviet community. At the same time, however, Central Asians cannot really separate themselves from Moscow, since the specialization that was once imposed on them will leave them dependent on the federation for a long time. These opposing tensions are exceedingly dangerous. They further fuel the bitterness of people at the periphery who are stubbornly attached to Islam and pitted against a repudiated center they still need for their survival. In Russia, these tensions produce fear, Islamophobia, and the urge to avoid a world seen as hostile and explosive. In fact, since everyone desires not to speed up the march to the "casket" and join the murdered immigrants, the whole European community is quite ready to grab their "suitcases."

The succession of riots and violent incidents that has been a constant fact of life in Central Asia since 1989 is surprisingly reminiscent of how the empire of czarist Russia went from slowly coming apart to exploding. As if by chance, Kazakhstan was prominent at both moments when this shock occurred; it was the nodal point in both cases. In 1916, when the empire had been weakened by the initial defeats in World War I and its subjects were distraught to find the colossus had feet of clay (the old version of society's disarray at the failure of the system), it was the periphery that first rejected the central government. On July 2, 1916, the nomadic Kazakhs rose up against the conscription imposed on them (another theme in the conflict of 1990), and revolts set the steppe ablaze. Besides the immediate reason for the revolt were the massive arrivals of Russian colonists, the battle over the land, and the fall in the standard of living. During this revolt of despair against the foreigners, thousands of Kirghiz and Kazakhs were massacred. The bloody repression never did reestablish calm but created a veritable "disorder front" extending from the steppes of Kazakhstan to the Transcaucasus. The Ferghana and the area of Samarkand were also almost as turbulent as the steppe. Everywhere, the people, alarmed at the dimensions of the repression and the plans for increased colonization, began mobilizing around a fanatical and conservative Muslim clergy who dreamed of a holy war. The promodernization intellectuals seeking their community's moral and political rebirth thought that Islam, but an renovated Islam, could muster popular energies to end imperial domination. The empire collapsed in this region before the revolution of 1917. Lenin realized this, offered a hand to all national aspirations, and at first marveled that in a rebellious Central Asia the mullahs would affix red stripes and the insignias of communism, sickle and hammer, to the green flag of Islam.

At the end of the 1980s, when memory had been restored to the Soviets, the peoples of Central Asia discovered this rebellious past, identified their current tragedy with it, and linked up with the combats of old. What is true of Central Asia is true of the rest of the violent periphery. The Caucasus, agitated by separatism once prohibited by the Soviet government, turned its gaze toward the legendary figure of the Imam Shamil, who had long respected Russian power. The Georgians remember the crushing of the revolt of 1924. From one republic to another, today's revolts seem to mirror past resistances. Glasnost, which gave each people the right to connect with its history, certainly contributed to reconnecting the past with the present, reminding the infuriated peoples that they had once before risen up against a powerful empire, even before it was defeated, and so had contributed decisively to its collapse. As soon as the truth, now revealed to everyone, showed the peoples of the USSR that they once again were headed for an abyss, past and present came together to forge in them the same conviction: their destiny must lie in their own hands.

Part Three

The End of Federalism

7

The "Double Power"

FOR nearly seventy years, Soviet life was governed by two principles: the Party was the sole possessor of power and the sole representative of the society (through the various social organizations it controlled); and any initiative, change, or reform could come only *from above*—from the Party. That is how perestroika began. Mikhail Gorbachev cleared the way for future changes by looking at reality and not myth and eliminating whole portions of the lie. Initially, however, the initiative was his, and the appeal for reform and the plans for achieving it came from above.

The remarkable aspect of Soviet history over the last five years is the way the people gradually claimed a role in the proposed changes and became the architects of reform. The Soviet Union experienced a veritable revolution, the first since 1917, which linked smoothly with the change announced in February of that year but interrupted by Lenin's coup d'état in October. Deprived of any initiative for seventy years, the Soviet people started regaining it in 1988. Pluralism flourished while the Party still claimed its controlling role, unwillingly giving it up in 1990,[1] when this role had clearly become nothing more than a fiction. The initiative and even the authority had shifted to grassroots organizations led by the people.

The source of this revolution was surely the people's desire to express themselves and participate in public life. Even before Gorbachev's ascension to power, special interest groups had emerged: volunteers gathered to discuss common concerns, chief among them salvaging the country's artistic and cultural heritage and defending the environment. After 1985, these groups rapidly multiplied, declaring that the people were not as resigned as they were reputed to be and would not forever be denied control over their own destinies. Gorbachev not only accepted but encouraged these forums for discussion, which were called *informal groups*. Perestroika needed society's mobilization to succeed. The emergence of these groups suggested that the limits of social apathy had been reached and that still-viable sectors of initiative and vitality could, by way of example, effectively contribute to the reforms' success. In 1987, Kurashvili, a lawyer who became an impassioned advocate of these grassroots initiatives, wrote, "General progress, reflecting the very essence of perestroika, lies in the radical advancement of societal self-management."[2]

The government acknowledged the need for the organizations that were springing up throughout the country but also felt that they had to be controlled—particularly because their dramatic proliferation caused uneasiness within the Party.

This was the source of the *popular fronts* created in part spontaneously, but in part encouraged, sometimes actively, by the government. The idea behind creation of the fronts was simple. Although mobilizing the people was indispensable for Gorbachev and for the Party, it could also prove risky. The Soviet government could not let these small groups proliferate indefinitely, for they embraced conflicting aims that, if unchecked, might eventually weaken rather than help it. By organizing these groups under the authority of broad popular fronts—which in the past had always been offshoots of the Party—the government hoped to win on two fronts—by securing the people's active support in exchange for tolerating shared initiative and by channelling this support and giving it direction.

Gorbachev soon realized how illusory his "frontist strategy" was. Born with the Party's blessing, the popular fronts quickly broke free of the Party and developed their own strategies, becoming the nuclei of genuine political parties. The popular fronts were assisted by a change in the electoral system and the establishment of a limited parliamentarism that soon outflanked its creators. With these reforms, the fronts were able to play an independent political game. They could not have existed without Mikhail Gorbachev and the reforms he decreed, but without these fronts and several other similar organizations, the impact of his reforms would have been reduced. The activities of these organizations—informal groups, popular fronts, and ultimately political parties—shattered the whole political system, imposing on it a de facto and, later, a de jure pluralism.

The creation of fronts and implementation of this revolution was particularly assisted by the framework of nation-republics. Underneath all their differences, the Soviet people harbored a common aspiration—to strengthen the union's component nations. As soon as it became possible to struggle openly for society's development, all problems—democracy, social progress, culture, ecology—were included in the same nationalist programs. This explains the speed with which popular fronts emerged in every republic, and their capacity to play a political role, since all of society, not just groups or factions, supported them. This led to an upheaval in the entire Soviet political landscape—even a challenge to the continued existence of the Soviet Union.

The Vanguard: The Baltic Popular Fronts

Battles to preserve national languages and impose their use,[3] battles for ecology, and justice for the system's victims were movements that fueled a profusion of organizations—the Greens, Memorial, Helsinki 86, and so on. Nearly everywhere, the vanguards of these battles were the national creative artists' unions (mainly writers and filmmakers). It was not by chance that in September 1988, the Odessa Film Festival concluded with a appeal to the whole Soviet elite to create a Popular Front for the Support of Perestroika.[4] At that moment, the two notions of popular front and the defense of perestroika were closely intertwined. This was the initial context for the creation of the fronts, and the first and surely most remarkable instance of the establishment of a front was set in the Baltic states.

Powerful popular fronts were formed in the three states annexed by Stalin in 1940, and from the start, they were representative. Within less than four months (June to October 1988) the scattered informal groups came together. It was not surprising that these three republics would be the first to set the example. Sovietized later than all the others, they remembered a politics not dominated by a single party. Still haunted by the memory of their lost independence, the peoples of the Baltic region were clearly better prepared than others to take advantage of perestroika to debate their own problems and aspirations.

It often has been objected that Baltic independence was short-lived—twenty-two years—and that the history of these states, once essentially annexed by Peter the Great, did not differ fundamentally from other Russian possessions. Nevertheless, the empire's "Baltic provinces"—"the open window on Europe"—long enjoyed a privileged status. Their cultural status enabled them to have prestigious universities in which the Russian language, which was imposed everywhere else, was not the rule (like the University of Dorpat, now Tartu). Their prestige was greater because they abolished serfdom in 1821, forty years before Russia. And although Alexander III tried to impose a policy of Russification on them, his efforts produced an explosion of nationalist feelings and a will to resist. This heritage was apparent as early as the spring of 1988.

The first popular front, that of Lithuania, known by the name Saiudis *(Lietuvos persitvarkymo saiudis)* was born on June 3, 1988, against the background of a national crisis. That date marked the forty-seventh anniversary of the massive deportations Stalin ordered after the reannexation of the Baltic states. Demonstrations were being readied in all three capitals, and popular agitation was mounting from day to day. This mood of resentment gave Saiudis a radical tone from the start. On June 24, the Lithuanian Popular Front organized a demonstration in Vilnius for the purpose of making its program public—support for perestroika, of course, but primarily a struggle for greater autonomy for the republic and for the defense of national interests in the areas of culture, language, and the environment. The newly born Popular Front emphasized its idea of perestroika: it must above all have a nationalist significance, without implying either chauvinism or any bypassing of democracy. But for Saiudis, the Lithuanian nation was a privileged framework for political progress. Within this framework and for these ends society could mobilize and help perestroika. We see that the fledgling front was beginning to move beyond Gorbachev's conception of reform.

The same vision prevailed with the creation of the Estonian Popular Front, which occurred in Tallin on October 1, 1988, during a large meeting of representatives from several informal organizations and associations from various republics. This rush of delegates to the fronts' founding congresses shows the extent to which the movement was spreading to the periphery and the speed with which liaison between the republics was recognized as valuable for their common interests—in other words, attaining greater independence from the center.

This development was confirmed in Riga on October 8, date of the birth of the Latvian Popular Front. More than a thousand delegates attended the founding congress—many of whom who earlier had fought successfully against the building of a dam on the Daugava River—and they already understood the strength generated by such gatherings. Along with individual members, the front included ecologists and members of various cultural associations, including Renaissance and Renewal, an association fighting for religious freedoms.

Front organization grew stronger from one congress to the next. In Riga, the immediate need for a structure and local contacts was acknowledged. By the end of a few weeks, the Latvian front could claim some 250,000 members and a considerable number of sympathizers. The three leaders of the Baltic fronts, remembering politics before Soviet annexation, knew that their survival depended on two factors—close contact with the people and good coordination among themselves.

Contact with the people meant control of the means of disseminating information, and all fought for access to radio, television, and printing presses. In this area, Saiudis was certainly the most successful, since it had a television program—"The Waves of Rebirth"—and created dozens of newspapers, including the legally recognized *Rebirth,* which had a press run of 100,000 copies, with one edition in Lithuanian and another in Russian. The privileged manager of this intense communication activity was a publishing cooperative. The Estonian front had access to television. In Latvia, however, the authorities at first allowed the front to speak out only occasionally. Even so, a weekly radio program and a weekly newspaper (in Russian and Latvian versions) with a press run of 100,000 copies regularly communicated with the public. In addition, the front lent its support to the publication of other organizations, particularly religious ones, thereby constantly broadening its influence.

These publications were thus of paramount importance; in each of the Baltic countries, they proliferated, disappeared, and were reborn under new titles. Although highly popular, they encountered everywhere a production problem—ever-inadequate allotments of paper. This may have been caused by a paper shortage—or the government, forced by popular pressure to sanction certain newspapers, may have surreptitiously tried to sabotage its concessions. What was noteworthy was the ability of independent publications to exist and maintain press runs unrelated to the amounts of paper allocated. The Soviet habits of counting on "parallel" supplies and on general resourcefulness surely account for this success.

A few short months after their creation, the problem of communication among the three fronts was resolved. Meeting in Estonia in May 1989, representatives from each republic created the Council of Baltic Popular Fronts to coordinate their activities, devise a common strategy to replace scattered activities, communicate with other political organizations in the region, and—not the least of its jobs—reflect on the future of the Baltic world.[5] This council often met to respond to the CPSU's mounting criticisms of the fronts' machinations and took over responsibility for the famous *human chain,* which is discussed later in this chapter.

The Baltic popular fronts had to face opposition from Moscow, the local Communist parties—heretofore masters of the political scene in these republics—and the large Russian communities residing in these countries.

At first Moscow appeared relatively indifferent to the innovation represented by these fronts, but when they claimed to be representatives of the various nations, the CPSU intervened to criticize positions that abandoned the Soviet institutional framework.[6] It sharply reminded the fronts that the Baltic countries were states of the USSR, where the controlling organ was the Party and the Party alone.

But the CPSU was counting on mainly its local contacts—the national Communist parties, to keep the fronts' activities within acceptable limits. The local parties were hesitant, however; they soon realized that the people supported the fronts and that, by clinging to an intransigent line, they ran a high risk of being totally rejected by a public that already saw them as tools of the central government.

Although the popular fronts had adopted a common strategy, the national Communist parties were far from doing so.

Apprehensive about being overrun by the local front, the Lithuanian party decided at its plenary session of October 20, 1988, to compete with the front in nationalist fervor. Its newly elected first secretary, Algirdas Brazoskas, accused the CPSU of meddling in the positions and activities of the republics' own Communist parties. His position wavered, however: he denied Moscow's right to appoint the second Party secretary, but he was forced to recognize that he himself had negotiated his own candidacy in Moscow.

In Estonia, the Party proved even more adaptable. It claimed to be cooperating with the local front for the success of its desired reforms.

At the other extreme, the Latvian Communist party opted for the hard line: it had no wish to cooperate with the local front and attempted to thwart all its initiatives. It also played the internationalist card against the front. The size of the Russian minority posed a considerable problem for Latvians (Latvia had 2,666,587 inhabitants in 1989, of whom 1,287,646 were Latvians, 905,515 Russians, 120,000 Belorussians, and 92,000 Ukrainians), who saw the Popular Front as a tool that would be used by people hoping to expel the foreigners from Latvia. The majority—probably even nearly all—of the Latvian people no longer accepted the foreign invasion that threatened their integrity, pushed for an industrialization that threatened to destroy the environment, and appropriated the best jobs for Russians. The Latvians saw their local Communist party, with its clearly pro-Russian stance, as a Russian organization.[7]

In this delicate context, the Latvian front sought to avoid any accusation of chauvinism, interethnic hostility, or anti-Russianism. Initially, it meant to deal with the problem of the balance among ethnic communities in measured terms—by ending immigration without repudiating immigrants already settled—but this relatively flexible position on the ethnic issue encountered three problems. Within the front, an ultranationalist mentality, determined primarily to fight against Russification and for the restoration of a Latvian identity without foreign influences, began to find a voice. The debate between the two sides—one more egalitarian, the other highly nationalistic—soon was won by the nationalists with the support of front members and the people at large.

The second problem concerned the emergence of mostly cultural nationalist organizations, many of which represented nationalities that had immigrated to Latvia—Russians, Belorussians, Poles, and Jews (due to emigration the Jewish population declined from 28,318 in 1979 to 22,897 in 1989). This proliferation of cultural groups led to a radicalization of nationalist attitudes, among them those of the Latvians themselves, which led to the possibility that the Popular Front would be overwhelmed if it did not appear sufficiently wedded to the defense of the Latvian cause. Finally, in January 1989, the Popular Front was joined by the newly created International Front of the Working People of Latvia (or Interfront), which was basically founded by Russians, had only a few Latvian members, and defended the idea of the "friendship of peoples," meaning absolute fidelity to Moscow. Not all Russians and other Slavs in Latvia belonged to Interfront, whose conservatism tended to discourage any advocate of reform. But the front tried to prevent the creation of this movement, as did some clear-thinking Russians, and appealed to the public in the monthly *Rodnik* (*Avots* in the Latvian version) during December 1988. The authors of this remarkable message stressed that Latvians and Russians could coexist only at the price of self-determination.[8] Nothing, however, could stem the growth of Interfront, and the Popular Front of Latvia was thus nudged in a more nationalistic direction than its supporters wished.

In the other Soviet republics, the example of the Baltics was followed attentively because the creation of their popular fronts had caused the three republics to become the chief rallying points for the non-Russian democratic organizations, from which all other fronts gradually arose. For the organizations that still hesitated about making the difficult transition from informal groups, which the government accepted, to a true Popular Front, which the government often opposed, nationalist debates proved to be invaluable: at these meetings groups that were still unsure of themselves discovered they were not isolated but belonged to an ever-growing network of organizations. They saw how their problems were shared and sometimes already partly resolved by other groups, as the Baltic fronts (in Lvov in June 1988, in Riga in September 1988, and in Vilnius in February 1989) showed the European Soviet peoples, who sent delegates from their democratic movements to these meetings.[9]

This regrouping of non-Russian democratic organizations became more radicalized from one meeting to the next. At first (in Tbilisi, Yerevan, and Lvov), they studied the utility of and methods for coordinating the activities of the various movements and electing general authorities. The meeting in Riga was largely devoted to the problem of perestroika and the forging of a common program of genuine decentralization in the USSR and economic and political autonomy for the republics. In Vilnius, the tone grew strident. The conference came up with two texts with instructive titles—"Charter for the Freedom of the Subservient Nations of the USSR" and "An Appeal to the Russian Intelligentsia." The authors of these documents dwelt on new topics—solidarity among oppressed peoples, both Russians and non-Russians; the tyranny endured by the Russian nation itself; and the delay of the Russian democratic movements in addressing nationalist issues. From this moment on, an idea began to take hold in the organizations of the periphery: since democracy required the solution of the nationalist question, the first priority for any authentically democratic movement was to struggle for the self-determination of the peoples of the USSR.[10]

Contagion

In a climate where the national elites were becoming aware of common problems and forces were beginning to act more or less in concert, the example of the Baltics very quickly proved contagious.

The republic where the government least expected this virus to spread was Moldavia. Appropriated from Romania during World War II[11] and forced to adopt the Cyrillic alphabet in order to cut its ties to Romania, Moldavia had long been a multiethnic republic with no apparent problems. In 1989, its 4,332,000 inhabitants included 2,790,769 Moldavians, more than 1 million Russians and Ukrainians (about equally divided), and a sizable Jewish community (65,668 in 1989, compared with 80,000 ten years earlier). The Soviet government had congratulated itself that 53 percent of the Moldavians spoke fluent Russian—that is, nearly as large a proportion as that of the Ukrainians and Belorussians—although their own language, Romanian, was unrelated to the Slavic languages. In 1988, a democratic movement erupted in this peaceful republic. It came from the depths of a society that suddenly allied itself with the developments in the Baltics and moved Moldavia into the vanguard.

At first, the Democratic Movement for Perestroika, founded on June 3, 1988, resembled other movements that were encouraged and even promoted by the government. It supported reforms and democratization—in short, Gorbachev's agenda. It also increasingly insisted on the autonomy of the republics, the control of immigration, and the appropriation of land by the still-numerous Moldavian peasantry. But two particular problems catapulted this movement's program beyond perestroika and transformed the movement into a popular front. First, the Democratic Movement saw language as a chief problem; it demanded that Moldavian become the state language of a republic that included numerous minorities whose cultural rights would consequently be diminished. Above all, they demanded that the Cyrillic alphabet be dropped in favor of the Latin alphabet, which would return the people to a linguistic unity that was lost in 1940. "Moldavian and Romanian are the same language split up arbitrarily. It is time to bring down the Great Wall of China separating us from our origins." This nostalgia for neighboring Romania also applied to the areas taken from Moldavia and incorporated into the Ukraine. Although the Democratic Movement did not demand the return of these areas, it still posed the problem and thereby worried Moscow, which refused to discuss the redrawing of borders.[12]

Other organizations were forming concurrently with this movement and often confused with it by their members—the Greens, cultural and scientific associations: all dealt with the same problem—the survival of a Moldavian nation that was threatened with destruction by the national policy of the USSR.

Despite the Democratic Movement's insistence on respect for the rights of all minorities, the latter were uneasy. The rise of Moldavian nationalism prompted Jews to depart in great numbers, and in January 1989 the Russians and Ukrainians created an Internationalist Front similar to the Latvian Interfront. This front demanded that Russian be recognized as the state language and equal to Moldavian and was violently opposed to the adoption of the Latin alphabet. It was, if not pushed, then at least supported, by Moscow and also the local Communist party, which was very close to the CPSU.

Since the founding of the Internationalist Front could in turn intensify the nationalistic determination and positions of the various movements, it is hardly surprising that their private conferences led to a meeting of all existing associations, in Kishinev on May 20, 1989, to found the Popular Front of Moldavia. Inspired by the example of the Baltic fronts, to which it was very close, and taking a cue from the reactions of the Russians in Moldavia and the central government's indecisiveness, the front openly proclaimed several radical demands: Moldavia must be fully sovereign; the problem of the lost territories must be a point of negotiations between the governments of the USSR, Moldavia, and the Ukraine, and the territorial status declared only temporary; all immigration into Moldavia and any attempt by Moscow to get Moldavians to emigrate from their republic must end because preservation of the ethnic group was an incontestable priority; the leadership must be replaced and free of any central control; the *nomenklatura* must be abolished; and the re-Latinized Moldavian language must be affirmed. All these demands were presented as a nonnegotiable package. From the start, the Popular Front raised the problem of the Hitler-Stalin pact and linked the fate of the Moldavians to that of the Baltic countries, improperly annexed after the signing of a treaty that was contrary to international law as well as contrary to the right of people to self-determination.

The front thus debated problems that challenged the USSR as presently constituted. Furthermore, the front's radicalism was not limited to words. The authorities' unwillingness to recognize the front immediately as it demanded led it to organize demonstrations that showed its representativeness and its popular support. Since the emergence of the democratic movement, these demonstrations, rallying thousands of people, made Kishinev one of the most turbulent cities in the Soviet Union. One day, crowds came out in support of perestroika; another day, they demanded the "rehabilitation" of Stalin's victims; a third day, they demanded that houses of worship be restored to the Orthodox Church of Moldavia. Ever more numerous, demonstrators held impressive marches to fight for the front's legalization, which they finally succeeded in obtaining on October 26, 1989. Everything in Moldavia became a test of strength, pitting the local government, which the central power obviously supported, against the republic's Moldavian population, which was united behind the front in ever more massive numbers. When the front was finally legalized, it was nearly indistinguishable from the Baltic fronts with which it continued to identify. Under the influence of the front, Moldavia had ceased to be a "model" republic.

Belorussia, another model republic although Slavic, also had a popular front, but its difficult circumstances required them to appeal for help from the national elites of Lithuania—further evidence of the active part the Baltic fronts took in the general developments at the periphery. The Popular Front of Belorussia was created in Minsk on October 19, 1988, and immediately encountered a hostile reaction from the local government. The Belorussian Communist party promptly declared that none of its members could join the front, which led many early enthusiasts to take a cautious stand. The first demonstration organized by the front, on October 30, was dispersed with unusual violence by the police.[13] The front appeared doomed: it was furiously hounded by the authorities who refused to grant it legal status or the right to demonstrate in the streets or publish or present its demands on radio and television.

Again, however, the example of the Baltic countries saved members of the front from despair. On June 29, 1989, many of them traveled to Vilnius, where they officially founded their popular front—this time with great ceremony and the participation of delegates from the Baltic fronts. Two births of a single front were a worrisome sign for Moscow[14] and, for the local authorities, an incitement to increase provocations by invoking the need to defend the rights of nationalities living in Belorussia (1,341,000 Russians, 290,000 Ukrainians, 418,000 Poles, and 118,000 Jews).[15] The Belorussians countered that between 1979 and 1989 a massive number of Russians and Ukrainians had settled in their republic, while the Belorussian population was stagnating compared to the rate of demographic increase in the Soviet Union (9 percent for the whole of the USSR, 4 percent for the Belorussians). This relative regression alarmed the republic's leaders. Interfront sent representatives to Belorussia to mobilize the minorities who were "threatened" by the local nationalism. Although the front did not have the right to hold meetings, these foreign arrivals met with no obstacles at the workplaces where a large number of non-Belorussians assembled.

After the front's second birth, its popularity (nearly 100,000 members claimed) could be explained primarily by the tragedy at Chernobyl. The government—both central and local—remained silent about the scope of the disaster, particularly the amount of radioactive land, after it delayed the population's evacuation from the contaminated areas. The front conducted parallel investigations into the amount of radioactivity and its consequences for the health of a population who never were informed of the dangers incurred at the time of the explosion and who had continued to live in the dangerous areas. Chernobyl was without any doubt the source of the growing estrangement between Belorussia and the USSR, which to that point had seemed perfectly attuned. The nuclear accident inspired the rapid creation and popularity of the front despite persecutions. In this republic, however, the Popular Front, symbol of national bitterness and anxiety, seemed to have a long way to go before it would be in the vanguard of political change.

In the Ukraine, on the other hand, where a front was late to emerge, the spread of reformist and nationalist ideas and the proliferation of the ongoing struggles made this republic a potential center for changes that were fraught with consequences for the USSR. The Ukraine was the second largest Soviet republic in population and economic importance. It had 51.5 million inhabitants, of whom 37 million were Ukrainian and 11 million Russian. Forty-four million Ukrainians lived in the USSR as a whole, which meant that after the Russians, also Slavs, they constituted the largest community within Soviet land area. A very large portion of the USSR's agriculture and industry was located in the Ukraine including linen, sugar beets, coal mines, steel mills, and a sizable number of nuclear power stations. Since Chernobyl, the Ukraine had refused to be the USSR's "nuclear waste dump." And it is still waiting for glasnost to produce an explanation of the disaster it suffered.

For years, the Ukrainian intelligentsia had raised the problem of threats to the republic's language and culture. All the evidence points to the reality of those threats. During the last decade, however, the Ukrainian people discovered that they were the victims of a veritable "demographic Chernobyl." While the population of the USSR increased by 9 percent, that of the Ukraine grew by only 3 percent, and the Ukrainian community itself by only 2 percent. In addition, in their own republic, only 87 percent of persons of Ukrainian nationality said that their native language was Ukrainian, while throughout the USSR, the percentage of persons of a particular nationality who claimed their national language as their native language was 92 percent. Nearly 60 percent of Ukrainians spoke Russian, however, and the nation's intellectuals were uneasy about the propensity of their people to adopt this language out of convenience or opportunism. This explains the important role writers played in the movement to create a popular front.

Although the Ukraine held a special place within the USSR, it also had other unique characteristics. First, the two parts of the republic were historically, religiously, and politically heterogeneous. The eastern Ukraine, long integrated into Russia, except for a brief period of independence, was always part of the USSR and indirectly shared in the world of Russian orthodoxy, including all of Soviet history for which it paid a frightful cost. Nowhere else was collectivization so ghastly: for Stalin, collectivization also helped stifle Ukrainian nationalism. He deliberately arranged for the great famine of 1932 and 1933 during which more than 3 million Ukrainians perished.[16] The war was just as ghastly, and the treason-obsessed Stalin once thought of deporting all the Ukrainians. He regretfully gave up the idea only because of the huge number of people involved.

The western Ukraine once belonged essentially to the Austro-Hungarian empire in which Ukrainian culture had flourished. Annexed to the USSR after World War II, western Ukraine had a different memory of the empire and also of the period between the two world wars. The population is largely Catholic, even though Stalin had forcibly annexed the Catholic to the Orthodox churches. The western Ukrainians, however, never accepted the loss of the Vatican's authority or that of their religious freedom. And in the 1980s, the Ukrainians began to entertain the hope that a "Slavic pope" would secure their return to the fold of Catholicism. They had benefited for several years from the courageous underground missionary work of the Polish priests who crossed the border in peril of their lives to minister to phantom parishes. The pope and the Polish clergy helped reopen, if as yet only imperceptibly, a border that since 1945 had been one of despair.

Although Gorbachev, who was anxious to eliminate all national leaders who served under Brezhnev, long retained the elderly Vladimir Shcherbitsky at his post, it was because the new General Secretary distrusted the reactions of Ukrainian nationalism that from now on would find nourishment everywhere.[17]

The Ukraine's delay in forming a popular front—which did not occur until 1989—stemmed in part from the numerous organizations (the Writers' Union, which was in the vanguard, the Academy of Sciences, the national society Memorial, divided into local and sometimes competing factions, the Democratic Union, the Helsinki groups, and so forth) that already flourished and in part from the dispersal of political activity between two centers, Kiev and Lvov, but especially from the local government's implacable opposition. Under Gorbachev, the local government stayed much the same as it had been under Brezhnev, because it had long been kept at the head of its Party and because the Ukraine had not been systematically victimized by purges after 1985. The numerous political demonstrations, rallying tens of thousands of Ukrainians, often were put down with extreme violence.

Despite these obstacles, the Popular Front for Perestroika (RUKH) was founded in Kiev on September 9 and 10, 1989, and immediately adopted a platform and elected the national and nationalist poet Ivan Drach as its head. Compared with the Baltic and Moldavian platforms, its initial platform seemed fairly moderate. It stated a desire for economic autonomy for the republic and focused on environmental problems, language, and, of course, human rights.[18] The document hardly presaged developments to come.

The founding of RUKH was a source of great perplexity for Shcherbitsky and Moscow. Unable to prevent it, the Communist leaders wavered between a relatively flexible attitude that they hoped would lead to the front's neutralization and a tough stance against "extremists who dream of seizing power, imposing on the people a nationalistic ideology, separation from the USSR, . . . [19] and a return to all the symbols of nationalism."[20]

One reason for RUKH's initial caution was that working out a common position for the two distinct parts of the Ukraine was extremely difficult. The informal organizations that grew up between 1986 and the founding of RUKH often held different if not conflicting positions on relations with the USSR and with Russia as well as on the tricky religious dispute. In this respect, creating a popular front for all of the Ukraine already represented considerable progress. Its leaders' primary concern clearly was to bring the opposing sides together, despite their differences.

In addition, the front was created during a time of political upheaval caused by a miners' strike in the summer of 1989. From this conflict, chiefly centered in the Ukraine, a powerful workers' movement emerged that spoke of its capacity for acting autonomously toward traditional authorities, local government, the Party, and the union. RUKH needed to form ties with this movement, a problem that was made all the more urgent since the Popular Front, where intellectuals had a leading influence, initially seemed to be dominated by the representatives of the western Ukraine, while the miners' movement primarily influenced the eastern part of the republic. Thus, relations with the miners had two goals—to unify the two parts of the Ukraine and to form a bond between the long-active intelligentsia and the hitherto silent working class.[21]

Next came the problem of RUKH's priorities: would they be national and cultural, as the intellectuals wished, or societal, as the miners wanted? These debates preoccupied the movement's leaders and gave temporary relief to Soviet authorities, who were well aware that the debates would delay the development of a popular front.

Despite these initial obstacles, the creation of RUKH was immediately seen, both inside and outside the USSR, as a turning point in the difficult history of the Ukraine. The presence at the founding congress of Lech Walesa's adviser, Adam Michnik, testified to the interest taken in this event by the Polish democratic movement and underscored its importance. Up to then, the creation of popular fronts had not received much publicity outside the USSR. With Adam Michnik, who brought the support of Solidarity and the Polish parliament to the Ukrainians and what he described as "demonstrations of national rebirth,"[22] the popular fronts of the USSR entered the great adventure of the democratization of Eastern and Central Europe.

Reconciling the Nation and Democracy

Among the belatedly formed popular fronts, the one in Georgia illustrates the internal difficulties experienced by the democratic movements at the periphery. By demonstrating in March 1956, Georgia had been in the vanguard of democratic rebirth but had lost the democratic habit. It regained it once again in 1978, on the occasion of a battle about revising the constitution of 1977.

Several characteristic elements of "the Georgian case" account for a development that, for a time, seemed to represent a step backward relative to the changes in, say, the Baltic countries. The brief but intense experience of democracy during the period of independence from 1918 to 1921 left the memory of a genuine political life, which explains the resurgence of a rash of small parties in the mid-1980s—a democratic phenomenon, certainly, but generating only scattered initiatives that tended to distract from effective regrouping. According to Georgian legend, wherever there are two Georgians, they immediately form two political parties—and splinter groups soon form in both of them.[23] A second element was the trauma of April 9, 1989, which radicalized and immediately mobilized public opinion behind demands for independence, before the popular front was prepared to assume full responsibility. Finally and primarily, the republic had serious ethnic problems: the front could try to rally all the groups to speed up the democratic process and give the front a broad foundation, or it could become strictly the defender of the Georgian community and risk clashing with other fronts founded by minorities, such as the highly active Popular Front of Southern Ossetia.

Georgia was not alone in having national divisions; the Baltic fronts also had to deal with them. In the Baltic states, however, the minorities, Russian or Ukrainian, had come from the outside and the debate did not concern national territory. In Georgia, at issue was the republic's territorial outlines and national structure—federal or narrowly ethnic. The real questions revolved around what Georgia should be at the close of the twentieth century and who was a member of the national community in Georgia.

Before a popular front appeared, the Georgian political scene was complex. Two groups, cultural in origin, argued about it in early 1988. Initially, the Ilya Chavchavadze (informal) association and the pan-Georgian Shota Rustaveli Society acted for the same causes—human rights, the restoration of former liberties, and the fight against cultural Russification. Public opinion, however, spontaneously mobilized behind more directly political views—the struggle against the constitutional amendments of 1988—and the two associations soon veered toward stronger nationalism and toward placing under their authority the many informal groups generated by political activists. Besides these two organizations, the Georgians could choose from more marginal ones, including the National Democrat Party (illegal, of course, since pluralism was not yet part of the Soviet system), the National Party for Independence, an association of Greens, and a Monarchist Party. The Writers' Union—here as elsewhere aware of the dangers of this fragmentation of positive forces—urged a regrouping—that is, a Popular Front, which emerged in July 1989.

Contrary to most of the informal groups, which lacked legal status, the front was immediately registered and hence integrated into the republic's political life. This rapid legalization sheds light on the original ambiguity of this gathering and many of its later difficulties. The intelligentsia wanted the front to heal rivalries among the informal organizations and provide a focus for growing popular enthusiasm. The republic's government—especially in the atmosphere of permanent crisis that existed after April 9—also supported the front, but as an instrument of control over the various existing organizations. The local government hoped to drain these organizations of members and thereby diminish their zeal. It also expected that the front's legality would preserve it, like its Baltic analogues, from overly extremist positions.

In fact, at its founding, the front's bylaws stated that it was a "popular universal sociopolitical organization uniting all people of good will to restore political, economic, and cultural independence to Georgia and create a democratic and humane society for all those who live in it." The front was weakened and considerable tensions were created by the inclusion of the theme of independence. The people wished to recognize only this demand, and if the front ignored them, it might risk isolating itself and missing an opportunity to plan for *all* the inhabitants of Georgia. In fact, this disunity within the organization did nothing to dampen popular demonstrations, and ever more radical demands came from the people.

Georgia then lacked anything comparable to the Karabakh Committee and its natural extension, the Armenian National Committee, founded in Yerevan on November 4, 1989, and immediately invested with a coherent civic and national platform. Although the committees held ambiguous ideas of sovereignty—for reasons connected with Armenia's special situation and its conception of sovereignty—this platform explicitly indicated the means for creating democracy and protecting the Armenian nation. There was no doubt about the legitimacy of the Armenian movement nor that it represented other similar political organizations. At its founding congress representatives of the popular fronts of Moscow, Leningrad, Latvia, and the Democratic Union of Hungarian Youth were present, evidence of the authority enjoyed by the Armenian assembly. The tragedy of Karabakh and the traditional cohesion of the Armenians here overcame the classic Georgian propensity for divisiveness.

In the Muslim world the popular fronts were deeply rooted in the soil of tradition and thereby left the mainstream. The two most active and innovative movements were the Popular Front of Azerbaijan, whose genesis owed much to the Karabakh issue, and the Popular Front of Uzbekistan.

In Azerbaijan, the front was officially formed in Baku only on July 16, 1989, but the Azeri intellectuals who promoted it had worked for months on a platform. Because of the Karabakh conflict and the subsequent military intervention of January 1990, the front could establish itself as a true alternative power and provide a focus for public opinion, which after the conflicts with Armenia, had taken a vehemently separatist stand in less than a month. The front was best able to devise a strategy for breaking away from the USSR and attaining Azerbaijani unity.

In Central Asia, the Popular Front of Uzbekistan, Birlik (Unity), showed the way for the still highly scattered informal movements in other republics. Birlik, a mass organization with more than a half million members, supported a series of large demonstrations (as in Tashkent on March 19, 1989) on behalf of its legalization, demanding among other things the elimination of the monoculture of cotton, a theme that crystallized all the Uzbeks' frustrations and provoked intense emotions. Birlik was unreservedly opposed to any economic organizing by the USSR, a position that suggests why its leaders did not bother to take cover behind the convenient label of support for perestroika. The only restructuring Birlik promoted was Uzbekistan's, an attitude that contributed to its popular success and to Uzbek mobilization. The rapid development of the political environment created by Birlik was reflected by an event whose significance escaped many observers—the founding in early 1990 of the ERK party, which is in all likelihood an offshoot of Birlik. First, this party is remarkable for its name, an allusion to the ephemeral National Party of Turkestan founded just after the revolution. It is also noteworthy because of the personality of its leaders, the chief one being the Uzbek poet Muhammad Solih, one of the leaders of the young intelligentsia who struggled to work out a compromise between national tradition and democracy.[24]

More than elsewhere, history weighs very heavily on the present in Georgia, Azerbaijan, and the Muslim states of the USSR. In their platforms and even their names, the popular fronts and informal groups constantly refer to the past as a period of grandeur that then serves as a symbolic reference point. The names of Rustaveli, Ali Shir Navai, or the Dzhadids, who inspired ERK, hover over these organizations and sometimes prevent them from clearly viewing problems. These fronts were in close contact with other organizations in various Soviet republics and tried, depending on circumstances, to either guide or follow a highly mobilized public opinion toward a new definition of the nation's interest and rights.

Symbolic Victories: The Chains of Unity

In August 1989 and January 1990, four popular fronts—Lithuanian, Estonian, Latvian, and Ukrainian—strikingly demonstrated their ability to marshal the people of their republics behind a unified viewpoint. The example came, as always, from the three Baltic states.

Once established, the Baltic fronts had to address a twofold problem that called for their full attention. First, they needed to offset the small populations of these republics (barely 6 million people) by uniting everyone in the struggle. The Baltic states had been so victimized by their past, which kept them apart, that the fronts did not try to resolve this problem. An initial response to this challenge was the creation of the Baltic Council of Popular Fronts. Beyond that, however, the Balts also meant to note the illegality of their annexation by commemorating in their fashion the Hitler-Stalin pact, the subject of broad debate now initiated by the popular fronts.

The Baltic states pressured the Soviet government officially to admit the existence of the secret protocols appended to the pact of August 23, 1939—so that they could denounce the protocols and the violation of international law represented by the annexations of 1940. They called for the creation in Moscow of a special committee from the Congress of the People's Deputies while an ad hoc committee from the Baltic parliaments simultaneously looked into the problem. These demands were made despite the Soviet government's early fierce denials of the existence of these protocols and then claims that the protocols could not be annulled without major risks to the balance of power in Europe.[25] But on the eve of the fiftieth anniversary of the pact, the fronts won the battle. Although Moscow did not yet explicitly recognize the need to invalidate the protocols and the consequences of the German-Soviet alliance, a large group of Soviet legislators did so, giving this act wide support.[26] On August 22, 1989, an official announcement denounced and invalidated the pact, not just in the Baltic states, but also in Warsaw, where the Politburo of the People's Republic of Poland published a bulletin condemning the treaty and its secret protocols as violating international law.

Against the background of this historical reevaluation—a moral victory in the hard-fought dispute with the Soviet government—the coordinators of the popular fronts thus produced another real triumph with the human chain. On August 23, the three Baltic capitals of Tallin, Riga, and Vilnius were linked by a human chain of some 2 to 3 million people (as always, exact estimates are hard to come by, and the Soviet media attempted to minimize the event) spread out over some 360 miles. At the borders, the leaders of each republic stood side by side and set off fireworks to celebrate this success marked with solemnity and joy. The dual symbolism of this demonstration was clear: the "enchained" people, the Balts, would proceed in unison along the road to emancipation. The reaction to this in Moscow was hostile, with the press calling the Balts "divisive forces" and enemies of renewal[27] or, more seriously, suggesting that the popular fronts were preparing for an armed uprising.[28] The overtly threatening tone of certain commentaries, and their recommendations for suppression, did little to intimidate the popular fronts. The human chain allowed the fronts to test their strength and their ability to nonviolently mobilize the people. Fortified by the recognition of their power, the Baltic national movements continued their peaceful march toward sovereignty. This huge demonstration accurately reflected the solidarity among three long-divided peoples. Some months later, this example was replicated by the Popular Front of the Ukraine.

For RUKH, the eternal problem was to rally the two parts of the republic and demonstrate that the nationalist demand was shared by all strata in the society, including the powerful miners' committees. One way to manage this was by getting inspiration from what the Baltic states had done. Here, also, there was remarkable success. On January 21, 1990—another symbolic date, for it marked the seventy-second anniversary of the founding of the Independent Republic of the Ukraine—a human chain some 300 miles long linking Kiev to Lvov, the capitals of the two parts of the Ukraine, was organized by RUKH and several other associations, a further illustration of the Ukrainians' aspirations for unity. Although it was difficult to estimate the exact number of participants, all reports agree that the chain was unbroken, that in some segments the demonstrators were so numerous that they formed three rows, and that in several cities this chain produced gatherings in the thousands and even tens of thousands. The number of participants was so great that many people who wanted to participate were forced to travel from their own cities to do so. The event's success increased RUKH's influence not only with the Ukrainians but also with minorities living in the republic.

Because of the importance of the Ukraine in the Soviet federation, the central government could not treat this event with the brutality displayed in the Baltics. The reaction in the Ukraine was cautious, all the more so because the movement's direction appeared to be undecided. Even though at the time their chain was formed, the Balts were already moving toward liberation from the USSR, the various groups within the RUKH, while determined to unite, had not yet spoken out on their ideas for the Ukraine's future: would it be inside or outside the USSR? Furthermore, the purpose of the Ukrainian chain was less radical, since it aimed to demonstrate Ukrainian unity, while the Baltic chain was meant to express a rejection of their annexation.

These two demonstrations were of inestimable political importance for Moscow. They had shown, possibly for the first time, that grassroots organizations could join together to mobilize the population of an entire republic. Up to then, these organizations had tested their capacities for mobilization only in limited areas. The human chains confirmed that the recently formed popular fronts had already achieved a maturity that enabled them to treat as equals the political authorities hitherto supposed to represent the national societies. The Communist parties, the unions, and all the social organizations issuing from the Party had just lost their legitimacy—or what was left of it.

Concrete Victories: The Electoral Battle

The political reform of December 1988 and the elections to the Congress of the People's Deputies gave the popular fronts a chance to measure the solidity of their access to the sphere of power. This was a formidable test, since the legislative elections of March 1989 were as yet only partly free, and the authorized plurality of the candidacies was inadequate to offset the prohibition still in force on political parties.[29] Moreover, the nationalist movements had to cope with two major problems. The first one stemmed from the legal status of these organizations. Although the electoral law of 1988 recognized every person's right to be a candidate in the election, and although the official social organizations had their quota of deputies, the popular fronts and informal groups fell outside the category of individuals and recognized social organizations and so in principle were not officially players in the election. The second problem arose out of the new distribution of deputies in the congress and in the Supreme Soviet, as enacted by article 18 of the electoral law and article 3 of the constitution revised on December 1, 1988.[30] The constitutional amendments notably decreased the number of representatives from the republics to the congress; deputies from three groups (representing the federation, the republics, and the social organizations) reduced this representation to a third of the whole, while in the former system, the two houses of the Supreme Soviet (Soviet of the Union and Soviet of the Nationalities) granted the republics half the representation. There were 750 deputies from the republics out of 1,500 deputies to the Supreme Soviet before the reform of 1988; after it, there were 750 out of 2,200. The same held true of the new Supreme Soviet of 1989, where the republics felt they were less well represented by fewer deputies. Consequently, the republics' electorate saw little reason to vote and required persuasive arguments from the popular front to be convinced that they should to go to the polls. Under these conditions, the election results, especially in the Baltic states, Moldavia, and the Ukraine, were all the more remarkable.

The electoral importance of the new organizations showed up most clearly in Lithuania. Fortified by its legal status and popular support, which had been affirmed under many circumstances, Saiudis had candidates everywhere running against those of the Party. There were only two exceptions to these contests: candidates were withdrawn in favor of A. Brazuskas and V. Berezov, first and second secretaries, respectively, of the Communist party of Lithuania. Because they had shown themselves relatively open to innovative ideas, Saiudis had feared, in the event of their defeat, that these two Party leaders would be replaced by two conservatives. This tactical retreat in no way diminished an electoral victory as stunning as it was unexpected. By the close of the balloting, Saiudis won 75 percent of the seats reserved for Lithuania, in both urban and rural areas. The successes of the first and second Party secretaries were counted among the seats won by Saiudis. Since Lithuanians made up some 80 percent of the republic's population, Saiudis must have received practically all the nationalist votes, and at the opening of the Congress on May 25, 1989, it thus was a real political party representing nearly all the Lithuanian people.

It was also remarkable that none of the candidates from the Russian-dominated Edinstvo (Unity) organization had won a single seat in the election. All the state and Party favorites had been beaten. Saiudis personified legitimacy. Nevertheless, the Popular Front was not in power. Just after the elections, a crucial problem arose: could national power and legitimacy be reconciled within the framework of the Soviet system?

In the other two Baltic republics, despite their large Russian electorates, the success of the candidates from the fronts or supported by the fronts was less sweeping but clearly indicated that a shift in legitimacy from government authorities to the fronts had occurred.

In Estonia, the Popular Front did not nominate any candidates but supported all Party candidates who belonged to the front and every candidate who upheld the front's platform. Of the twenty-one candidates backed by the front, eighteen were elected. Since the republic had the right to thirty-six deputies, the front had earned a respectable victory: these eighteen deputies—that is, 50 percent of those elected—closely corresponded to the Estonian share of the population, which was around 60 percent. As required by the ethnic composition of the population, seven candidates from Interfront were elected in the area that had a predominantly Russian population.

In Latvia, the Popular Front found the election results no less gratifying. The republic had a quota of forty deputies. As in Estonia, rather than nominate its own candidates, the Popular Front supported the reformist candidates who explicitly backed its platform. The influence of a non-Latvian population that approached half the total population probably played a decisive role in this choice. But the Popular Front lost nothing: half the winners in the voting were members of the front, 70 percent supported its ideas, and Interfront trailed far behind.

In Moldavia, the electoral victories of the nationalist candidates were even more surprising, for everything was working toward their defeat: the conservative government apparatus was supported by a Communist party openly hostile to perestroika; the government pressured, often nearly threatened, the candidates and the electorate to preserve the status quo; and the campaign was nondemocratic, in total contradiction to the provisions of the 1988 electoral law guaranteeing the candidates equitable treatment by the media and equal access to meeting places. Acknowledging their inability to overcome these obstacles on a nationwide scale, Moldavian nationalist organizations chose to run in a limited number of electoral districts—sixteen, just less than a third, which they won by a majority. Including all the Moldavian deputies—those elected and the representatives of social organizations—the reformers obtained close to a third of the fifty-five seats.

Strengthened by their incontestable legitimacy, reformist deputies from Moldavia to the Congress actively participated in all the debates on national problems. In particular, they demanded parliamentary investigations of the repressions conducted in Georgia and in Kishinev, political control of the police and legislation regulating their activity in the republics, and a new linguistic policy consistent with the platform of the popular fronts. Thus, in the parliament, the elected Moldavian nationalists tended to become the spokespersons for all nationalist organizations.

These unexpectedly positive results, attained in an appalling electoral climate, considerably strengthened the position of Moldavia's nationalist organizations and forced the central government to attend to their demands. At the same time, the apprehension expressed by the minority groups at these victories captured attention in Moscow, where a "minorities strategy" was being worked out against the growing nationalisms. This dangerous strategy risked provoking or inflaming interethnic clashes that Moscow seemed at a loss to control.

The elections in the Ukraine were marked by the resounding defeat of many Party and state leaders. Shcherbitsky—in the plenary session of the Central Committee of the Ukrainian Communist party, which convened to discuss the election disaster—pointed to the activity of the Popular Front and various informal groups that had mobilized the society against the Party as having won negative votes from the electorate. The defeats suffered by the Ukrainian *nomenklatura* were one side of the coin; the other was the affront represented by the election of several candidates who would become the future leaders of RUKH or other nationalist organizations. No doubt the elections of 1989 were premature in the complicated context of the Ukraine where the more than 3 million members of the Communist party constituted an impressive force capable of putting a whole roster of undesirable candidates to rout. No nationalist organization had been able to establish itself sufficiently to mount a real political campaign (RUKH was still seeking to do so). Personalities from this nascent nationalist movement, however, won; the informal groups' harassment of the Party was enough to win popular support for reformism.

The Ukrainian Communist party took a dim view of these unexpected results. Just after the session, Shcherbitsky proclaimed, with highly undemocratic animation, that glasnost "leads to extremism and a demagogic nationalism." His virulent speech called for the Party's veritable mobilization against a policy that it had never accepted. But in the Ukraine, as elsewhere, however, it was too late to check the course of democratization. The creation of RUKH some months later showed that a hard line had no effect on people who were awakening to political life and to ideas worth striving for.

The other movements in the republics did not win as dazzlingly in the elections because throughout the Caucasus the rise in interethnic tragedies and suppression of agitation created an unpropitious climate for the smooth unfolding of an electoral campaign. In Central Asia, the small amount of activity within the popular fronts was a result of pressure from the *nomenklatura* and the lack of political traditions. In Kazakhstan, all the regional secretaries of the Parties were the lone candidates in their voting districts, a fairly common situation in this part of the periphery. Elections there were pointless, but their lesson was not lost: the popular fronts concluded that it was important to prepare for the local elections of 1990, since the interests at stake were vital to the electorate and should be taken seriously.

At the beginning of 1989, however, the popular fronts focused their efforts on the problem of language. They also were concerned with ecology, an area in which the very active Greens threatened to diminish their following.

Victory for Tomorrow: Language

The first effect of glasnost was that all or nearly all the periphery threw itself headlong into the linguistic battle. Ukrainians, Georgians, and Uzbeks complained that the bilingualism so ardently desired by Moscow as a tangible sign of the existence of a Soviet people resulted in a loss of national identity and the slow deterioration of the national language. Gorbachev's openness induced the non-Russian nations to give this problem the highest priority; all the fronts and informal groups seized on it to mobilize the people behind a straightforward demand and use the plan to lead the people in a radical shift of direction.

The linguistic battle actually involved two distinct problems. The titular people of the republics wanted their language to take precedence over both Russian and the languages of their minorities when the latter had also acquired rights. To justify this, the large nations intended to use *legality*—the Soviet constitution and specific Soviet laws. When the USSR called itself a *state of law,* legal guarantees became paramount.

The second and more complex aspect of the battle over languages concerned alphabets. National identity implies not just adherence to the national language, but also an attachment to a traditional alphabet linking it to the wider cultural group to which the nation belongs. The nationalist revolts were to tear to pieces the whole Stalinist policy of linguistic unification through the universal adoption of the Cyrillic alphabet.

Until the mid-1980s, the most intransigent nationalists left the job of guaranteeing the status of their language to the republics' constitutions. The three Caucasian republics had long ago written into their constitutions that their languages were the official state languages. In 1979, during the constitutional revision, this article had been retained following impressive demonstrations. In the 1980s, however, a general nationalist awakening in the USSR had indicated how inadequate the constitutional guarantees were. Some of the smaller groups demanded that these guarantees be applied to themselves, while the Russian language imposed itself in practice owing to the silence of the constitutional texts on these smaller languages.

After 1985—with glasnost and perestroika promoting change—the national states reopened the issue and this time gave their languages a legal status that ensured their survival.

The three Baltic peoples were, of course, in the front ranks of this fight, but two other pioneers were Moldavia and Uzbekistan, which were the last two countries to wage the twin battles of legal status and the alphabet.

The laws were not all identical, however, and reflected a greater or lesser flexibility toward the central government. The latter was more distressed by what constituted a major international defeat for the USSR than by its moral and political effects.

Estonia worked out the most drastic statutes in this legal marathon.[31] Its law recognized only Estonian and granted no privileges to Russian. Every one of the republic's residents had four years in which to learn

Estonian, and everything written in Russian must immediately vanish from all public places. Lithuania,[32] although it had a smaller Russian population, proved to be more elastic about the prescribed time periods, since it allowed for a transitional period of two years before the same arrangements went into effect. Latvia, whose population was nearly 50 percent immigrants, needed to be more accommodating by leaving each of the republic's residents free to choose either Latvian or Russian for use in public life. Russian terminology, however, also had to disappear.[33]

In this revolution of interethnic relations, Moldavia, however, proved the most resolute in refusing to make any concessions.

The Writers' Union and the Popular Front united to argue for a plan to make Moldavian the official state language and to limit the use of Russian to relations with the rest of the Soviet Union. Supported by the Party, the Supreme Soviet presented a plan in which Moldavian was certainly recognized as the state language but in which Russian would be the vehicle for the *internal* communication between all the nationalities of the republic. The Writers' Union and the Popular Front organized demonstrations in Kishinev (close to a half million demonstrators on August 28 and 29, 1989) to show that the popular will was on their side. In the face of this, the Russian community did not remain dormant: the local soviets and the Party organizations from the cities with large Russian populations (Tiraspol, Rybnitsa) bombarded the local government with motions condemning the Moldavian plans. On August 28 and 29, a Russian coordinating committee, Soiuz (Union), and the Edinstvo (Unity) association in turn mobilized the Russians. While the Supreme Soviet debated the problem, a general strike paralyzed Tiraspol: the strike committees assembled tens of thousands of demonstrators demanding that Russian be given an equal legal footing with Moldavian and that the alphabet issue (a return to the Latin alphabet) be subjected to a referendum. Mikhail Gorbachev made his personal contribution to the interethnic conflict in the debate: he used all his authority to press the republic's first secretary, Grossiu, to have Russian named the language of interethnic communication.

For four days, the Supreme Soviet debated under conflicting pressures from the Moldavians and the Russians and under heavy pressure from Gorbachev. The final vote marked a triumph for nationalist demands. Moldavian was recognized as the state language, the language of interethnic communication, and the Cyrillic alphabet was replaced by the Latin one. With, however, a slight bow toward Moscow: Russian (although not solely) was *also* recognized as a language of interethnic relations.[34] For the firmly committed Gorbachev, this failure was bitter.

Following the example of the Moldavians, the Uzbeks also waged a dual fight for their language and the restoration of the Arabic alphabet. Legally, they had been preceded by the Tadzhiks, who had ratified the official status of their language on June 22, 1989, but the alphabet problem remained unresolved. Of course, like the Moldavians, Muslims had begun, well before the law authorized them to, to disseminate their alphabet through new publications or newspapers that suddenly began using the two alphabets simultaneously. Throughout the Central Asian republics, this "pedagogy of the alphabet" had gained ground, and courses in "reading Arabic" were begun. Here, again, the popular fronts orchestrated a change in which Uzbekistan was in the forefront. The terms of the debate were same as those that had governed the adoption of the Moldavian law: what should be done about Russian? As far as the alphabet was concerned, the cause had, in fact, already been won. In Uzbekistan, however, Russians were in a position of retreat: they were preparing to leave, and many of those who had to remain behind began familiarizing themselves with the national language.

The mounting linguistic revolution in the USSR had many consequences. The debate had fueled the conflict between communities, and the victory of national languages left Russian communities in the republics and the Russian people in general even more embittered than before. This rejection of their language exposed the frailty of the "friendship of the peoples." Big Brother was naked.

This victory also affected other people within the USSR. Morale was lowered, of course, by this encouragement of nationalism. In addition, people united by a common alphabet that is incomprehensible to others will tend to form a community turned in on itself. There was an international effect as well, as certain borders of the USSR thus acquired a linguistic permeability, bringing together groups that previously had been separated. By readopting Latin script, the Moldavians thereby rejoined the Romanian community.[35] By regaining their own alphabet, the Azeris united, as they wished, with their conationals in Iran. This was true of all the periphery, except for the Georgians and Armenians, whom Stalin had allowed to retain their alphabets and who lived basically within their own borders.

• • •

The popular fronts and movements were decisive players in the breakup of the empire. By mobilizing national populations behind aspirations or latent anxieties that were finally expressible because of glasnost, these organizations became genuinely representative. The legitimacy they thus obtained allowed them at the very least to exert pressure on the local government and even Moscow and, by invoking the popular will they embodied, to guide or reorient official policy. Most significantly, they circumvented the single-party system still in effect, entered the political area, and competed victoriously with the Party, whose collapse they signaled.

In the space of two years, during which pluralism was not allowed and the empire was officially intact, these representatives of the people—unevenly, but just about everywhere—helped bring about the birth of a participatory civil society, the establishment of a de facto pluralism, and the transformation and even disintegration of the empire.

It is hardly surprising that men like Yegor Ligachev, wedded to the idea of maintaining the Soviet system, observed with horror that the flowering of informal or recognized organizations led to the establishment of a dual power[36] *(dvoevlastie)*—which implied that the Soviet system that had ruled for seventy years had virtually ceased to exist.

8

From Sovereignty to Independence

SOVIET federalism rested on several pillars. The first was the law. The Constitution of the USSR applied universally, and the constitutions of the republics had to be consistent with it (article 73 of the Constitution of 1977). Article 74 specified that the "laws of the USSR have equal power over the territory of all the federated republics. In the case of a discrepancy with the federal law, the latter is to take precedence."

The second pillar was the army: "Service in the armed forces of the USSR must be a positive school for internationalism."[1] To carry out this purpose, the military law of 1938, which abolished national military units, stipulated that military service was to be carried out in ethnically heterogeneous units outside the national territory.[2] The Soviet Army used only one language, Russian.

The final pillar was the Communist party, "the force that orients and guides Soviet society" (article 6 of the Constitution of 1977). The Party was the symbol of the unity of Soviet society. All the republics had their own local parties, but these parties were merely representative of the Communist Party of the Soviet Union, as indicated by article 41 of the Party's statutes adopted at its Twenty-seventh Congress in 1986.[3]

Winning Sovereignty

Rediscovering the historical truth, if only imcompletely, played a major role in reaffirming the sovereignty of the republics. Nevertheless, the meaning of *sovereignty* is rather imprecise,[4] for it covers both the Soviet practice of federalism—"the federated republic . . . is a sovereign state" (article 76 of the Constitution of 1977)—and the most extreme demands for independence and secession from the USSR.

When the Baltic states decided to proclaim their sovereignty, their decision acknowledged two truths. First, Soviet-style sovereignty was merely an illusion (the three "pillars" mentioned above stripped it of any content), and the Soviet-style "free association," the foundation of the whole system (article 70 of the Constitution), had never existed in the Baltic countries. Second, the questionable legality of the Hitler-Stalin Pact,[5] which led to the annexation of the Baltic countries, made it possible for them to claim sovereignty: their annexation had been on the basis of an agreement so iniquitous that it had to be kept secret, which gave the Baltic countries a special right to assert their sovereignty. The legalism of the Balts, the full significance of which was not appreciated by Moscow, was at the heart of the three republics' claim first to sovereignty and then independence—that is, secession.

Thus, in less than a year—November 1988 to July 1989—the three Baltic states affirmed their sovereignty and made legal arrangements for translating it into reality. Although the idea of sovereignty had been in the air before 1988, the popular fronts made the word their platform.

The Constitutional Aspect

Estonia, which already had a radical language policy, was also the most radical in its will to regain its lost sovereignty. On November 16, 1988, the republic's Supreme Soviet announced its sovereignty, and its parliament immediately voted for constitutional amendments that would give the proclamation some substance. On May 18, 1989, Lithuania did the same followed by Latvia on July 28.[6] From that moment on, the three Baltic states formed a unique entity within the Soviet Union. This distinct position was strengthened by the decisions of the Council of Popular Fronts, which started integrating the three republics economically. At no time between asserting sovereignty and forming a common union did they consult Moscow. Moscow was faced with a fait accompli.

The first change to be made following this unilaterally proclaimed sovereignty was in relations between the USSR and the republics; a treaty of international law was required between the states. In other words, the USSR had to negotiate with the republics about their role in the federation. The second basic change was in the primacy of laws of the USSR and the republics. Soviet law would apply only when consistent with and approved by the republic's supreme soviet (Estonia, article 74 of the amended constitution; Lithuania, article 70). Latvia proved more cautious on this point: in article 71 of its revised constitution, it affirmed the primacy of the republic's law but also that of Soviet law in the event of conflict.

The republics also expressed sovereignty by appropriating all natural resources and by claiming an exclusive right to decide their own policy in every area, notably economics. The three republics affirmed their economic autonomy by adopting laws just after their declaration of sovereignty that committed them to radical reforms.

Sovereignty also affected electoral law and future local balloting. Always one step ahead of the others, Estonia wished, by a new ruling about voting privileges, to curtail the rights of residents who were not Estonian. On August 8, the parliament enacted legislation specifying that, to become eligible to vote, every non-Estonian Soviet had to meet a residency requirement of at least five years or an uninterrupted stay of two years prior to the elections. The general outcry over these provisions by the Russian community—their strikes and demonstrations paralyzed the whole country—and the Supreme Soviet of the USSR's vigorous condemnation of this arrangement for noncompliance with the Soviet Constitution,[7] led the Estonian parliament to retreat, but only for a limited time. It voted to suspend but not to cancel these provisions for the local elections of 1990.[8] This gave Moscow some temporary satisfaction, but left the problem unresolved. In fact, it was postponed until the framing of the citizenship law, where the most radical ideas could reemerge. The platform drawn up in July 1988 by the Latvian Popular Front stated that citizenship would be granted only to individuals who had resided in the republic for at least ten years. The rapid march toward independence, however, made this debate pointless.

The central government had various reactions to this rejection of federal supremacy. Moscow greeted Estonia's declaration of sovereignty with hostility and battled with it over the electoral law. The mobilization of the Russians in Estonia was also a sign of the apprehension generated by this declaration. When Latvia declared its sovereignty, however, Moscow seemed to reconcile itself to what it took as simple formal declarations, and it tried to regain the initiative in economic matters. On July 27, 1989, the Soviet parliament (Supreme Soviet) passed regulations that established—or, rather, recognized—the economic autonomy of the three republics and confirmed their transition to an "accounting economy" *(khozraschet)*. Moscow realized the need to come to terms with the fait accompli,[9] and it tried to take advantage of the situation by persuading the Baltic republics to serve as guinea pigs for the implementation of reforms that Gorbachev planned to impose on the whole USSR. A success would benefit everyone, and a failure could also teach a lesson.

But the example of the Baltic countries proved contagious. First, it was not just a matter of pure form. Lithuania brazenly vetoed various regulations (notably rejecting taxation on income from state enterprises and unilaterally deciding the total amount of pensions). Similarly, Estonia adopted a price and tax policy independent of the federation's policies. On January 16, Alexandra Biryukova, the Soviet vice prime minister, protested that economic autonomy had to respect Soviet laws. No one in the Baltic states seemed to care about these objections.

The Baltic example proved most irresistible in the Caucasus, where it seemed to indicate a potential response to local conflicts. The law declaring sovereignty that the Supreme Soviet of Azerbaijan voted for in September of 1989 implied a territorial concept of sovereignty. Azerbaijan claimed that it controlled its own political organizations and that only the republic's soviet or a popular referendum (a clear allusion to Nagorno-Karabakh) could reassign that control. As in the Baltic countries, the Azeris' declaration of sovereignty stipulated that Soviet law applied to their republic only when consistent with its own law. Finally, the exercise of the right to secede depended on a popular referendum within the republic. Azerbaijan also kept the exclusive right to proclaim a state of emergency. These provisions limited the central authority by barring it from any territorial adjustments.[10] Moscow disputed its constitutionality but also ignored it in practice. This law represented merely a first step, however, for at the start of 1990, the republic's Supreme Soviet announced that since they could not impose the law's sovereignty, they would sponsor a referendum on the question of secession.

Armenia also proclaimed its sovereignty, calling itself the United Republic of Armenia, and Georgia did the same in a more radical way.

In Georgia, however, this affirmation of genuine sovereignty immediately seemed inadequate, for two reasons. The shock of April 9 united the people in an anti-Soviet bitterness that time did nothing to attenuate but in fact sharply escalated. The second reason stemmed from the escalating demands in the republic for separation—during the summer of 1989, the Azeris in Georgia were pressuring for territorial autonomy in the areas south of Tbilisi, where they had settled in great numbers—that convinced the nationalist movements that Georgia was threatened with a Moscow-devised dismantling in order to crush nationalist sentiments definitively. At the time, all the plans presented in Tbilisi were dominated by the idea that the very survival of the nation of Georgia was at stake. From this viewpoint, sovereignty, which left some initiative to the Soviet federation, was to be discarded in favor of total independence. As for principle, the Georgians were the first in the USSR to mention the right to secede and connect it to the conditions of their participation in the union—annexation through military force of 1921, despite a friendship treaty and an internationally recognized status of independence.

In June 1989, Georgia's Writers' Union called on their Supreme Soviet to proclaim genuine sovereignty (similar to that of Estonia) with the immediate creation of Georgian citizenship and, if possible, to organize a popular referendum on self-determination.[11] On March 25, 1990, nearly 100 political parties and organizations of all kinds that had begun to prepare for independence in Georgia (this diversity within the nationalist movement indicates the Georgians' problems in agreeing on the exact steps to take toward sovereignty) nevertheless managed to gather together in a National Forum for Independence. The assembly, comparable to the Supreme Soviet, was to lead negotiations for separating from Moscow. At the same time, the republic's Supreme Soviet amended the constitution by introducing a multiparty system and decided to postpone local elections—then being held throughout the USSR—to the fall in order to make them "national" elections organizing the independent republic.[12] The Georgians thus took the route of the Baltic countries, but with a difference. By changing the electoral calendar and especially by giving a quite different form to the local elections of 1990, they took up a position outside Soviet life and clearly indicated that Tbilisi and no longer Moscow was the legitimate capital.

Another newcomer to sovereignty was Uzbekistan, which on June 20,

1990, adopted the Estonia model of 1989, although the Uzbeks' were reluctant to stop after such a good start rather than go all the way by proclaiming their independence.

This was also the position of Moldavia, where rising separatist demands made the local Communist party suddenly switch from rejecting any idea of sovereignty to adopting a plan of support of political[13] and economic sovereignty within a transformed federation. A month later, on June 23, 1990, the republic's Supreme Soviet voted for sovereignty as well as for the appropriation of all natural resources; the republic declared itself a demilitarized zone, rejected the principle of dual citizenship (Moldavian and Soviet), and announced that it would seek admission to the United Nations.

Finally, the July 1990 proclamation of the Ukraine's sovereignty in highly radical terms—the appropriation of all the republic's natural resources, a decision to form a national army and print its own money, and the adoption of a Ukrainian citizenship—showed that the spread of this phenomenon to all of the USSR was draining the federation of any real substance.

Russia's eventual proclamation of its own sovereignty constitutes a quite different story, one which we shall return to further on. Nevertheless, it too demonstrates that sovereignty, contrary to all earlier Soviet practice, had in one year overrun the entire union, thus rendering Soviet federalism null and void. By its bold move, Estonia had in one blow shattered the whole system, which Moscow now had to rethink from top to bottom.

The Military Aspect

The republics' desire for sovereignty caused the Soviet Army—the "army of the friendship of the people of the USSR"—to suffer as much as the constitution. Both within it and in public debate, the future of a multiethnic army was more than just cast in doubt. Within the army, it was clear that, far from representing an integrating influence on the nationalities, the army was one of the hot spots of interethnic hatred and violence. Particularly alarming was an official report by the military justice authorities that estimated that 20 percent of the crimes committed by members of the army and from 40 to 70 percent of the gross infractions of discipline were "overtly ethnic or racial in nature" and that "a great number of military personnel convicted of crimes are of Central Asian and Caucasian origin."[14] This report was confirmed by the army general N. Popov, who wrote, "The nationalist elements play on popular emotions and tend to aggravate interethnic antagonisms. These elements are disrupting the young people. Some of them introduce into their units nationalist and religious acts and prejudices that are alien to our society."[15]

These phrases reveal a serious dispute that exploded when glasnost made it possible to express all frustrations. Conscripts from Central Asia complained that they were reviled and treated like slaves by the Russians, some of whose remarks confirmed the charge:[16] "From the start, we, the whites, considered ourselves better than the *Churki* (Central Asians). If we had something nasty to do—like cleaning the latrines —we had a Kazakh do it. In the barracks, anything repellent was assigned to the Kazakhs and the Uzbeks." But the Balts were equally victimized by brutality in a military service in which physical and psychological violence had become the rule.[17] In the army, pitched battles took place between the "whites" and the Muslims, as well as between Muslims from different nations; in the navy, clashes occurred among Caucasians who were divided on the question of Nagorno-Karabakh.[18]

In short, any reason was valid for introducing into such ethnically diverse units the interethnic violence that from then on divided the Soviet people. The hazing of new recruits by more senior soldiers—the *dedovshchina,* a phenomenon that the press condemned as another sign of the decay of Soviet society—was in principle merely an expression of differences in military seniority;[19] but countless personal reports published by the press at the periphery revealed that the *dedovshchina* increasingly tended to pit the bullies against the bullied by national groupings, which increased the hazing and often verged on racial violence, pure and simple. This violence increasingly resulted in deaths that glasnost now allowed to be made public: a Baltic soldier accused some Russian soldiers of raping him and killed them, and, in return, the Balt was killed;[20] Uzbeks were murdered or rumored to have been murdered. Nothing was missing from this tragic picture. It must be added that officers were mostly Russian, since the government had failed to recruit more officers from the peripheral populations, who were incompetent or little tempted by military life,[21] which added to a climate of conflict that military leaders made no attempt to dispel.

Non-Russians felt that the army was a hostile setting, so it is unsurprising that the nations rejected it. Any pretext was valid for evading military service. Conscripts did not report to military authorities, who were alarmed at the growing number of desertions,[22] countless declarations of conscientious objection (even though these were disallowed), and hunger strikes, supported by popular demonstrations, by recruits refusing to serve in the Soviet Army (these had become common, notably in Georgia). The rapidly spreading hostility to the Soviet Army headed the list of the national movements' demands. From the still temperate demand for a change in Soviet military rules, the movements often went on to a more radical demand: "The army of occupation must leave our soil."[23]

In this regard, the departure of the Warsaw Pact troops from Hungary and Czechoslovakia, which was negotiated in 1990, had an extraordinary accelerative effect in the USSR, notably in the Baltic states where since its inception Saiudis had been demanding a return to national units and the establishment by each republic of its own officer-training schools to oversee these units. As soon as the Baltic states proclaimed their sovereignty, they initiated a study to reform the military, and, as an inevitable result, the young conscripts deserted in ever-larger numbers, expecting to serve in the army of their own republic. On the day of its founding, the Moldavian Popular Front's platform announced, as a corollary to sovereignty, the formation of homogeneous military units, even though it still admitted that units remaining within the Soviet Army could be stationed outside the republic. The Supreme Soviet of Moldavia, however, said that it would decide where their units would be stationed and that the principle of national homogeneity was nonnegotiable. The Popular Front of Georgia was no less intransigent and, while waiting for a change in the Soviet military service, asked all its sympathizers to protect the conscripts who refused to answer the draft. The front said that this was not desertion but a legitimate position of waiting. The sovereign Ukraine proposed a radical solution to the problem: it intended to have its own army.

Squeezed from all sides and aware of the discontent surrounding the army—the attacks on *dedovshchina* and the violence and criminality in the military—the Soviet government tried to defuse this bomb somewhat by making some concessions to the most turbulent nationalities.

The prime minister of Estonia was the first to apply to the Soviet minister of defense to use the maximum possible number of recruits as border guards and troops of the ministry of the interior. He procured an ambiguous agreement that was periodically criticized in Moscow but still enabled the national government to protect many young men from conscription. In Lithuania, the Party's first secretary, Algirdas Brazoskas, ever careful not to distance himself too much from public opinion by way of Saiudis, pleaded his cause with General Yazov. Here, too, a secret agreement was reached noting the fait accompli, even though General Yazov continued to oppose the formation of national units. The Georgians drew an argument from these precedents, which were supposed to remain secret but were well known, that married recruits should benefit from the same privileges in their republic, which could well produce an increase in the marriage rate.[24]

These local arrangements were much less significant, however, than the general debate in which they figured. The Soviet military leaders, chief among them the minister of defense, on the whole came out in favor of keeping the existing system.[25] The Party supported them. In June 1988, the Twenty-seventh Congress of the Party stated outright the need for retaining the status quo. On closer examination, we note that the army was divided on this point, and some of its leaders considered that a return to national units, far from having an disintegrative effect, might be helpful in remedying the worst ills of an army undermined by national conflicts. Thus, the commander of the Baltic military district, Lieutenant-general Kuzmin—elected to the Congress of the Peoples' Deputies—suggested that to avoid the total disintegration of the Soviet Army in the general hostility, the solution of national units might constitute a lesser evil.[26]

Soviet leaders also expressed apprehension at these growing ethnic disturbances within the army during the Party's 1989 plenary session on national problems. The minister of defense once again attacked the "nationalists, separatists, and extremists who are undermining the army of the interior." It was clear that this was also the Soviet government's position.

The desire of the nations to reappropriate the armed forces reflected two concerns. For the national leaders, a desire to assemble the elements of real sovereignty was accompanied by a genuine fear at the growing risk of the military consequences of a national conflict. Everyone dreaded that someday, as in Tbilisi, their republic would experience a violent repression and that a contingent of troops sent to reestablish order would include recruits from the same republic. Everywhere, there was a fear of what would then resemble a civil war: the nationalist movements were prepared for any sacrifice but did not want their compatriots killing each other. Avoiding this prospect by keeping their young conscripts within its borders, the republic would thus assure itself of a means of defense in the event of a military repression.

This debate about the army remained somewhat theoretical in Moscow, where the national recruits were not trusted. In addition, the overall reorganization of the army—as a professional army of volunteers or the draft—was debated. In this context, the issue of a return to national units was of only secondary importance. But this debate was crucial at the periphery for beyond the basic problems, any local government that decided to protect its absentees must have wondered whether this might not send the Soviet military leaders into a violent reaction.

Indeed, for these leaders, the debate about principles in the organization of military service was one thing. It was quite another to witness the wave of absences without leave, which were all the more worrisome because of the substantial contagion among Russian recruits. For everyone, the Soviet Army, now discredited in the press by exposures of *dedovshchina,* appeared more and more like a hell that no one was in any hurry to visit.[27] Beyond the relations between the nationalities and the army, the growing number of absences without leave (7,000 cases in 1989) helped to alienate the army from the Soviet people.

Breaking with the CPSU

Until recently, the CPSU remained the final pillar of integration in a crumbling USSR. But confidence in it as representative of the people's interest collapsed, for a poll conducted in December 1989 indicated that only 4 percent of the population still gave it any credence.[28] Despite this estrangement from the people, it was still within the Party that on September 19, 1989—nearly three years after the riots in Alma-Ata—Gorbachev launched the first large-scale discussion of the nationalist problem.[29]

At the close of his long speech—whose length heightened the importance of his words—Gorbachev expounded on the problem of the relations between the central Party and the national parties, a problem nearly everywhere already resolved by the dominance of the popular fronts over the parties. Gorbachev put forward some contradictory statements. He said that an authentic democratic centralism headed in the right direction must give the republics' parties more freedom, consistent, of course, with the general line of the CPSU. At the same time, however (and here he proved firmer), Gorbachev came out against the federalization of the CPSU—that is, against genuine independence for the national parties. "That would be the end of our Party, which was founded by Lenin!" he declared. And it is true that in his time Lenin had vigorously led a battle against attempts to federalize the Party and had won it.[30]

Nearly eighty years later, on July 10, 1989, his disciple Gorbachev was to lose this battle. The Lithuanian Communist party—which since Saiudis's electoral victory survived only by gambling on nationalism—adopted a new platform stating that the party was the champion of the "independence of the Lithuanian state"; and to bolster its new image, the party announced a plan to hold a congress where it would decide on its own independence. This episode was of considerable importance, for it marked the end of Moscow's authority over a national Communist party and underscored Gorbachev's personal lack of authority outside the walls of the Kremlin.

For some months, both Gorbachev and the Central Committee of the CPSU put great pressure on the Lithuanian Communist party. Alternately pleading and threatening, Gorbachev had many contacts with the first secretary of the Lithuanian party Brazoskas, to persuade him to proceed no further along the road to rupture. On November 16 the whole leadership of the Lithuanian party was summoned to Moscow to hear that no party could leave the CPSU. Two weeks later, Vladimir Medvedev, the Politburo's chief ideologue, arrived in Vilnius, bearing a message from Gorbachev. It was not impossible to ignore all these warnings—especially for the head of the party in a country with less than 4 million inhabitants. The memory of Tbilisi was not so distant that these repeated exhortations and their threatening overtones could be ignored.

Nevertheless, meeting in a special session on December 20, 1989, the Lithuanian Communist party unhesitatingly voted to break with the CPSU. It was the first secession ever to occur in the USSR, and it was by a Communist party. The breakdown in the voting is instructive: out of 1,038 delegates, there were only 183 negative votes. This figure must, once again, be seen in light of Lithuania's demographic makeup—80 percent Lithuanians. It was clear that all the Lithuanians at the Communist congress had voted in favor of the break.[31]

Gorbachev reacted as violently as Ligachev to this stunning result. Both mentioned "the general anxiety of the communists" and the "Lithuanians' ignorance of the difficulties awaiting them," words reminiscent of the words usually spoken on the eve of a brutal reaction by the USSR, whether against its own citizens or those of Eastern Europe.

What was unheard-of in this crisis was, first, the tranquil courage of the Lithuanians. There was no hesitation, as shown by the vote and the subsequent lack of schism (between Lithuanians) within the independence party. Also new was the cool indifference with which they greeted the pressures from Moscow. From then on, Gorbachev clearly had no authority over the Lithuanian Communist party, nor over Lithuania in general; Moscow no longer wielded the power of decision. Another element of surprise was that, save for the virulent words from the CPSU's leadership in Moscow, there were no other reactions, and this absence helped to strengthen the nationalists' certitude that Gorbachev was in a position of weakness in the face of their will for sovereignty.

Why then didn't other national parties follow the Lithuanian Communist party along the road of secession? In all likelihood because, compared to the popular fronts, these parties did not have enough prestige to engage in the solitary battles they feared they would not win. Moreover, no sooner had the Lithuanian Communist party set the example of independence than history accelerated. People proceeded from the independence of the party to the independence of the state. Sovereignty no longer sufficed for the pace-setting republics. Here again, Lithuania spearheaded the movement.

David against Goliath

Until the secession of the Lithuanian Communist party, Gorbachev thought he could withstand the independence trends, but after December 20, 1989, he realized the need to avert events before they occurred rather than acting only as they were going on or even afterward. On January 11, 1990, he went to Lithuania, preceded by "ambassadors" from the CPSU. During this extraordinary visit (no Russian leader had set foot on Lithuanian soil since 1945), Gorbachev and his colleagues spared no effort in trying to persuade the Lithuanians to relinquish their dream of separation and, instead, to wait for the effects of perestroika to improve the lot of the whole USSR. Gorbachev cited every argument possible—especially fear for the future of perestroika, which Lithuania's behavior threatened and the economic interest of Lithuania, whose development and organization were closely linked to the USSR. Breaking these links would impoverish Lithuania; it would not replace them with links to the capitalist countries as a new market for its products. "Independence means paying the prices of the world market, and you will sink!" prophesied Gorbachev to a crowd listening in frosty silence. This threat of economic disaster was accompanied by a bit of thinly veiled blackmail. The USSR had constructed ports in Lithuania that were needed by the Soviet community as a whole but that supplied the republic with considerable revenue in hard currency. Lithuanian independence would cause immense harm to Soviet military and naval installations as well as to individuals who would not wish to remain in an ultranationalist state. In the event of separation, all these losses would be subject to compensation.

Even more explicitly than Gorbachev, an article published at this time estimated the value of the ports,[32] and asked whether the USSR could surrender its investments to Lithuanian self-interest, or whether Lithuania was accountable for what the USSR had sacrificed for it.

Gorbachev's speech was explicit about economic pressure and the debt that would need to be repaid as a result of secession. Besides brandishing a heavy stick, however, he added a carrot that his interlocutors did not find very appetizing: the USSR was to become a "true federation," which he admitted it had never been, and laws would be enacted governing the right to secession, which would allow the Lithuanians to act, if they still wished to, legally.[33]

Politely listened to by various audiences, Gorbachev clearly made little impression on either the leaders or the crowds. The idea of sovereignty had been eclipsed by the dream of independence—*nezavisimost*. The local elections of the spring of 1990 helped Saiudis to win a great victory on the platform of independence for a Lithuanian state, neutral and completely separate from the USSR.

These elections were of major significance. First, they exemplified a real confrontation between already highly organized political parties. One Communist party was independent, the other had ties to the CPSU; one was a social-democratic party; another, a Christian-democratic party; and yet another, an ecological party. For the most part, Saiudis, which was not strictly a political party, supported independent candidates or candidates from other parties who backed its platform. The independent Communist party made every effort to defend its positions. Brazoskas opted for a platform similar to that of Saiudis, with a few differences in nuance: Lithuania would achieve complete independence but remain a Soviet republic with a special status. Nevertheless, Saiudis won a crushing victory, and the few winning communists were voted for because Saiudis had supported them.

The republic's Supreme Soviet, born of genuinely free and pluralist universal suffrage, thus had an indisputable mandate—to bring about full independence. Meeting on March 11, 1990, it voted almost unanimously (124 affirmative votes and 6 abstentions) for the restoration of independence, reaffirming the validity of the vote of 1918 and securing historical continuity with the Lithuanian state that had been independent until 1940.

Also unanimous were the votes to change the republic's name, which ceased to be "Soviet" or "Socialist" and reverted to the Republic of Lithuania; to transform the Supreme Soviet to the Supreme Council; and to elect Vitautas Landsbergis to the presidency. The new president announced that he was going to ask Moscow to recognize Lithuania's recovered independence, and he declared that any Soviet demand for financial reparations would in turn be met with demands for reparations to Lithuania. To emphasize independence, the Supreme Council's first act was the adoption on March 13, 1990—the day of Gorbachev's election to the presidency of the USSR—of a law freeing Lithuanians from service in the Soviet Army. The situation could not be clearer: the law supported Lithuania's decision, and Moscow could only ratify it.

The Soviet reaction was marked more by distress than by a clear grasp of the event. "Illegal," "worrisome," "inadmissible"—Gorbachev, Ligachev, and all the other Soviet leaders used the same vocabulary in reacting to the declaration of independence, even if Ligachev promptly took the precaution of announcing a ban on the use of force. For Gorbachev, however, there was nothing to negotiate because the USSR negotiated only with foreign powers, which in his eyes Lithuania was not: it was still a part of the USSR.

On March 15, the Congress of the People's Deputies showed its support for Gorbachev when an overwhelming majority of them refused to recognize the validity of the declaration of independence (1,463 votes for the motion, 94 against, and 128 abstentions). Lithuania appeared extremely isolated and guilty of having blundered. Gorbachev then issued an ultimatum: Lithuania had three days to abide by the position taken in the Congress of the People's Deputies and to recognize the illegality of its declaration. From that point, the possibility could be envisaged, if Lithuania asked for it by referendum, of starting negotiations on the long process of secession, the procedures of which remained to be defined.

In this confrontation between Moscow and Vilnius, begun on March 11, Moscow was mainly fearful about what would happen next. Would the Lithuanians' resolution to become independent—which had been relatively cautious the year before—promptly spread to Estonia, which had spearheaded these movements, and then to Latvia? How could Lithuania be isolated? How could people who were tempted to follow its example be frightened off and deterred? Urgent questions, for though the two other Baltic states were behaving cautiously, their future direction left little room for doubt.

Estonia took an especially original tack. Following the election of the spring 1990, the Estonians set up two legislatures: the Supreme Soviet was elected within the framework of Soviet laws; and the Congress of Estonia was elected by committees of citizens. In fact, however, the Congress of Estonia was immediately seen as the legal parliament of the future republic. Although illegal by Soviet law, the congress nevertheless was immediately accepted by the Supreme Soviet of the USSR. Among its members were the major players in Estonian politics, most of them Estonians elected to the Congress of the People's Deputies, like Marilu Lauristin, the president of the Popular Front, and many others.

The Congress of Estonia declared the uninterrupted existence of the Estonian state (since a large number of countries had never recognized its annexation). The congressional program was to restore the republic on the basis of the treaty signed in Tartu in 1920 with the young Soviet state led by Lenin. Gorbachev claimed to bear the mantle of Lenin; let him prove it by carrying out his forerunner's treaties. On March 12, the Congress of Estonia sent to the Congress of the People's Deputies a statement demanding the republic's de facto restoration and the immediate opening of negotiations to settle all problems connected with its recovered independence. The republic's Supreme Soviet followed suit, creating a commission to negotiate independence from Moscow and differing from the congress on only one quite theoretical point: whether there had been an interruption in the life of the republic. Was there any continuity between the state of 1918 and the independent state of 1990 exercising its sovereignty over the same territory and the same population?

The republic's Supreme Soviet quickly forgot these few points of discord and eventually sided with the congress's stand. On March 30, it declared by seventy-three votes to three abstentions that the Estonian state had never ceased its de jure existence. According to the congress, the country's status had been that of an "occupied state" and added that a proclamation of independence was invalid, since it amounted to giving legitimacy to a government of occupation. With this declaration, the republic's Supreme Soviet also began a transitional period that would last until the institutions of the independent state could function normally. Although this was not a declaration of formal independence of the Lithuanian type, the spirit behind it was the same. Moreover, the total alignment of the republic's Supreme Soviet and the congress reflected the triumph of institutions that were already post-Soviet. The Estonian government also asserted that it had no further military obligations toward the USSR and guaranteed the protection of any of its soldiers who were absent without leave. Thus, the constitutions of the USSR and Estonia were in open conflict.

Despite the handicap of a large Russian population, Latvia entered the game. On February 15, 1990, its Supreme Soviet voted by 177 to 48 a declaration that was more balanced than the other two but still affirmed the need to restore an independent Latvian state.

Thus, three distinct positions existed, and Lithuania was ahead of its two neighbors. In all the Baltic states, however, the declaration of independence was the result of an explicit popular mandate (as in Lithuania and Estonia) or a parliamentary vote with a wide majority (in Latvia). The question arose whether there was yet time for Moscow to intimidate the Estonians and the Latvians by threatening Lithuania and whether a threat could make Lithuania back down.

Goliath Entangled

To counter Lithuanian independence, Moscow deployed a huge arsenal of Soviet law, economic and military pressures, and the threat of territorial amputation.

The Weapon of Soviet Law

During his first visit to Lithuania, Gorbachev announced the drafting of a law on secession. The agenda of the Congress of the People's Deputies, however, confirmed that this law was to be debated only at the end of the session. The March 11 announcement made passage of this law urgent. The congress immediately seized on the plan and on April 4, 1990, voted for a law specifying the procedures for secession.[34] Fortified with this document, Gorbachev told the Lithuanians that they were in conflict with the law, since the law of April 4 did not provide for any unilateral declaration of independence. The Lithuanians were thus asked to return to square one, renounce their declaration of March 11, and begin the long journey provided for by the law.

Not content to use this law on secession against Lithuania, the Soviet government invoked international law—notably the Vienna Convention of April 23, 1978, to which the USSR was a signatory. According to this convention, a self-determining state must respect the laws in effect in the state to which it belonged but wished to leave—a proposition that enabled Moscow to brandish the new law of April 4![35]

The Lithuanian response was swift and double-pronged. First, Lithuania affirmed that secession was applicable (in the very terms of the constitution) only to countries that had *freely* joined the USSR, which was untrue in this case. Since it was a country annexed by force, like its two Baltic neighbors, Lithuania did not come under the category of the states described by article 70 of the Soviet Constitution. Furthermore, Lithuania objected that its declaration of independence was consistent with article 72 of this constitution—"Each republic preserves the right freely to separate from the USSR"—and did not contravene the law on secession, which had not existed on March 11, 1990. Internationally, Lithuania's position was all the more incontrovertible because most of the leading nations had never recognized its takeover by the USSR, which French president Giscard d'Estaing recalled in 1975 on the signing of the Helsinki final accords.

In 1990, most of the countries of the West (with the United States and France in the lead) cautiously adopted this position. Logically enough, they responded to Lithuania's demand for recognition of its sovereignty by saying they could not recognize an independence that they had always considered legally valid, even though it had been de facto eliminated. Although Lithuania was on more solid legal ground than Moscow, force was not on its side, which explains why, in refusing to waste time in juridical quibbling, Gorbachev favored pressuring the renegades.

Economic Pressure

Economic pressures were easy to apply. Economic interdependence was the rule in the USSR, and Moscow could strangle the republic by an expedient blockade, even though Gorbachev immediately stressed that he did not intend to reduce the people to famine. The blockade, mainly of energy sources and raw materials, considerably affected the republic's productive capacity and daily life. Moscow also threatened to block the export of goods from Lithuania to anywhere in the USSR, which would immediately affect the republic's finances.[36] Nonetheless, the blockade was not totally effective. Lithuanians also provided the USSR with consumer goods, particularly foodstuffs, and in the general scarcity, their absence could only aggravate the effects of the economic crisis.

For the blockade to be successful, the USSR also had to ensure that it was observed by all of Vilnius's trading partners (particularly neighboring Estonia, Latvia, and Belorussia), through which petroleum products came on their way to Lithuania and the outside world. Lithuania also had to be deprived of revenues in hard currency earned from traffic through its ports.

Moscow set about effectively discouraging any foreign aid (at first Poland judged it prudent not to make a move, as did the Western nations), denying ships and tourists access to the coasts and the republic's borders. Historical experience testifies, however, that these measures have but limited impact, especially when they fail to intimidate the countries that endure them. What was remarkable was the Lithuanians' determination and evident indifference to the material problems caused by the blockade. In this respect the people's solidarity with the independent government was reminiscent of the mobilization of the British during the Battle of Britain in 1940.

Military Pressures

Military pressures are harder to take because it is unclear whether the pressure is a matter of simple intimidation or will lead to total military intervention. After March 11, Moscow blew hot and cold. After its political leaders first declared that they would not resort to force, Marshal Akhromeyev, Gorbachev's adviser for military affairs, took the opposite position: "I am not afraid of using force if necessary," he declared.[37] The threat was all the more serious since the Soviet troops did not have to travel far to get to Lithuania: they were already there, as were the troops of the MVD. They performed some spectacular maneuvers—holding parades of armored vehicles in the streets of Vilnius, taking over some public buildings, and, above all, conducting brutal operations to abduct absentee soldiers from the hospitals where they were being sheltered. Like the blockade, however, the military threat was never completely enforced. Occasionally, a Soviet leader—like Gorbachev at the summit in

Ottawa on May 30, for example—indicated that he was ready to use force. Because they raised this threat too often without putting it into practice, however, the Soviet leaders also raised doubts. The Lithuanians, who had made the rather theoretical policy decision not to appear frightened, gradually concluded that they had no reason to be so and that these threats were meant to preserve Soviet dignity as much as to convince the republic to submit.

The Territorial Threat

In the last analysis, the weapon of territorial partition was perhaps the most worrisome, for it opened the door to future conflicts between neighboring nations.

On March 11, two territorial problems were addressed in Moscow—the future of Kaliningrad (formerly Königsberg), Russian access to which was challenged by Lithuanian independence, and especially the fate of Vilnius and Klaipeda, which the USSR had attached to Lithuania in 1940 at the time of the country's annexation.[38] At issue here were areas that in 1939 were under Polish sovereignty and that, when the USSR annexed western Belorussia in 1945, had, for historical reasons, to be incorporated in the Belorussian republic. The Lithuanian-Belorussian border, drawn in 1940 and finalized at the war's end, incorporated these districts and essentially Vilno (or Wilna), which became Vilnius, within the Lithuanian republic. Until 1990, no one in Belorussia realized that there was a disputable issue here, and Moscow can be suspected of provoking the suddenly fiery Belorussian claims to these territories in Lithuania. These claims were made in an official demand by the Supreme Soviet of Belorussia, addressed to Vilnius, to renegotiate its frontier. For Belorussians, it was self-evident that if Lithuanians left the USSR because they rejected the consequences of the agreements of 1939 and 1940, Belorussians likewise had no reason to feel bound by the accords that deprived them of a part of their territory.

Could Lithuania survive with losing Vilnius, several districts, and the port of Klaipeda or with being split by a corridor connecting Russia with Kaliningrad? The Lithuanian leaders dismissed these territorial threats without further discussion. Nonetheless, they understood they could not treat them casually, for on this point, the law was not entirely on their side.

As soon as it made a resolute start along the road to independence opened by Lithuania, Estonia became the object of similar threats. Interfront, which rallied the Estonian Russians, announced that in the event of a formal secession, the Russians would in turn demand the right to separate from Estonia and make the northeastern part of the country, where they were settled, a region attached to the USSR. Later, they added the region of Tallin to it. Even though Estonia could not seriously envisage thus losing a part of its territory, its leaders greeted this plan calmly, for they suspected it stemmed from indirect pressure from Moscow rather than a serious threat originating in the Russian community.[39]

Toward Secession

Two months later, despite all warnings, Estonia and Latvia caught up with Lithuania. Following the cautious approach that guided all its actions in this matter, Estonia was content to stockpile decisions strengthening its declaration of March 30—refusal to fulfill military obligations; refusal to contribute to the USSR's budget for defense and the maintenance of order (meaning the KGB); establishment by the law of May 16 of a transitional period during which Soviet law and legal procedures ceased to apply. Symbolically, after reassuming the republic's traditional name, Estonia toppled the statues of Lenin. Finally, after pro-Soviet communists and provocateurs tried to seize government headquarters, the authorities mobilized a national militia—largely made up of soldiers absent without leave from the Soviet Army or potential conscripts—and charged it with maintaining public order and, if need be, defending the country.

Latvia set out to be more formal. On May 4, 1990, its Supreme Soviet proclaimed by law the independence of the republic. The president of Lithuania had been invited to this formal session, and his presence symbolized the Baltic solidarity behind these decisions.[40]

Like Estonia, Lithuania mobilized young people to form a national guard, a potential army that could not be very effective against a Soviet military intervention but that testified, like the Estonian guard, to the republic's determination to defend its independence by every means.

On May 12, the presidents of the three countries met in Tallin and reinstated the old prewar Council of Baltic States,[41] which was empowered to arrange for the transition from formal independence to real independence and to promote cooperation among the three states. The primary purpose for resuscitating the council was to allow the reborn Baltic states to present a united front to the outside world. The council asked Gorbachev to open negotiations for normalizing relations between the three states and the USSR and for settling the problems connected with the separation. It also asked the signatories of the final Helsinki accord first to accept the Baltics and then to serve as a framework for their reestablished independence.

Although ignored by the signatories of the final Helsinki accord, the Baltics' actions did not pass unnoticed in Moscow. Gorbachev was no more amenable in May than he had been in March to accepting unilaterally decided independence—or that, at least, is what he said. He denounced the Estonian and Latvian behavior as unconstitutional and told the republics' lawmakers to respect the Soviet constitution and federal laws. Accompanied by the insinuation that Latvia and Estonia also could be slapped with economic sanctions, these provisions were initially hard to interpret. It was unclear whether they were genuine threats or a gambit to gain some time in the hope of negotiations that would save face on both sides.

Moscow's first expression of its official position was probably genuine verbal intransigence. As he had done since March 11, Gorbachev maintained that the Baltic states existed illegally, that their status was an internal affair in the USSR, and that it was not in any way open to negotiation, since that applied to relations between states and not to internal relations within a state. If the Baltic states renounced their independence and worked within the framework of the USSR and its constitution, a compromise might be reached.

As radical as this position was, it still created some suspicion. Although he was prodding the Baltics to bow to Moscow's might, Gorbachev was already half-opening hundreds of doors. He would open even more after the Soviet-American summit. Moreover, this chronology is easily understandable: Gorbachev could not meet with President Bush after making concessions to the Balts, for fear of seeming to yield beforehand to the Americans' "concerns." No one would go to a summit in a position of extreme weakness, having already given away his strongest cards. Following the meeting, however, fortified by the conviction that no official pressure would be exerted on him on behalf of the Baltic countries, he was free to prepare for appeasement.

In Moscow on June 12, he received the three presidents of the Baltic Supreme Soviets and suggested that they "freeze" their declarations of independence to allow for the opening of negotiations. On the surface, nothing had changed, but this was nevertheless a considerable step. First, to freeze these declarations was not to renounce them. They continued to exist, even if, for a transition period, the Baltic governments were committed not to take steps contrary to Soviet law. In addition, Gorbachev adopted a flexible position on the procedures for seceding, and he admitted that it could be quicker than the the laws indicated. Anatoly Gorbunov, the president of the Supreme Latvian Council, was probably right when he noted that Gorbachev's proposal amounted to an implicit recognition of independence. One can freeze only what already exists. And the hitherto unyielding Lithuanian leaders were probably right to accept the principle of this freeze as soon as the USSR halted the economic blockade.

The way was clear for a negotiated solution—not of the "Baltic crisis" but of the secession of the Baltic states. David and Goliath were apparently equal. But the Baltic David had had such a poor start three months earlier that simple survival constituted a undeniable victory. Nevertheless, Moscow was motivated to avoid prolonging a conflict whose consequences for the rest of the federation were already disastrous.

The Baltic Model

On March 11, Lithuania's declaration of independence was greeted with skepticism in all the republics, which called the spectacular gesture of the Parliament of Vilnius foolhardy. In Moscow the law regulating secession was passed by a large majority, and that law forced Lithuania to back down, attesting to the initial solitary position of the Baltics. Even more, when the Estonians and Latvians hesitated to follow the Lithuanians, they led Moscow to believe that by adroitly manipulating threats and peace offers it could avoid the creation of a unified Baltic front.

But this miscalculation soon became apparent. Estonians and Latvians eventually dared to speak out, however, even at the risk of appearing divided and hence vulnerable; after a brief hesitation, they made many gestures of solidarity with Vilnius. Similar expressions of support came from the rest of the USSR where the popular fronts (such as RUKH) but also official authorities (the Soviet of Moscow, for example) organized demonstrations on Lithuania's behalf. Particularly active was the Popular Front of Belorussia, whose position opposed its own republic's government and the territorial claims on Lithuania.

The Baltic countries were indebted to the Belorussian front for rallying the popular fronts on their behalf and for working out two plans that also had Moscow worried—"a political union of the western republics of the USSR extending from Estonia to Moldavia" and a Common Market covering the same territory.[42]

The Muslim states, too, expressed solidarity with demonstrations and impressive acts of support organized by the popular fronts. In Uzbekistan, Birlik sent an open letter of support for the Lithuanian cause to the Soviet leaders, the leaders of all the republics, and the United Nations. All the informal groups and political parties of Azerbaijan met and proposed that the republics help Lithuania "pay its ransom to the USSR," if it was forced to, and to set up diplomatic contacts with the newly independent republic.[43]

Moscow could live with this show of support as long as it came from political organizations with an uncertain political status. But government and official authorities, too, were increasingly supporting the fronts. Through the vehicle of the Supreme Soviets created by the elections of 1990, Moldavia, Belorussia, and Georgia formally recognized the right of the Baltics to self-determination and the legality of the procedures they had used, since their independence had been declared by freely elected parliaments.

On May 31, Gavril Popov, the new mayor of Moscow, also voted into office during recent elections, made a decisive gesture by announcing that he would establish direct trade relations with Lithuania for the purpose of supplying goods to the capital of the republic. In Leningrad, on June 26, Anatoly Sobchak, a newly elected mayor in Leningrad, entered into political and economic talks with the Estonian government. Finally, as soon as Boris Yeltsin was elected president of the Supreme Soviet of Russia, he announced his intention of signing a treaty of cooperation with the Baltic states.

In relations with the outside world, despite widespread caution, the icy noose around the neck of the independent states began to melt. Poland supplied Lithuania with raw materials, and Vaclav Havel, the new president of Czechoslovakia, received President Landsbergis in Prague.

Even though François Mitterand and Helmut Kohl then called on Lithuania to negotiate, their message clearly urged the two parties to enter a reasonable dialogue. This message also implicitly recognized that the conflict between Moscow and the Baltic states was of an international nature. This was exactly what Vilnius was claiming and formed the basis of its appeal for negotiations concerning Lithuanian independence. Finally, when the Council of Baltic States addressed the Helsinki signatories and the United Nations, it became evident that both—especially the Helsinki signatories—could not ignore these states for long. The isolation of the Baltics was very close to being a memory, and Moscow already was on the defensive.

The "Baltic model" not only prevailed over the skeptics but also attracted disciples. In its march toward secession, Georgia adopted the argument used by the Baltics, thereby hoping to take an "international" road to independence instead of going through the difficult procedures of secession. Georgia claimed that its legal government had never relinquished its functions and had never freely joined the Soviet Union. Suddenly, secession was of no more concern to it than to Estonia. Although it is unlikely that Moscow felt that the Georgian case was comparable to that of the Baltics—freely or not, Georgia had signed the pact of union that created the USSR on December 30, 1922—this development was bound to make Moscow uneasy, for it showed that the republics would use anything inside and outside the USSR as an argument for their own cause.

The events that were shaking Eastern Europe—frontiers fractured or moved, people who rediscovered themselves —resonated extraordinarily at the Soviet periphery, which encouraged the march toward independence. The Moldavians posed a serious problem for Moscow in this regard. The uncertain nature of Romanian politics would be an incentive to remain patient for a time, but the popular demonstrations at the border and the small groups that cut down the barbed-wire fences separating Moldavia from Romania (hadn't the revolution in Eastern Europe begun when the barbed-wire fences between Austria and Hungary were dismantled?) showed that the idea of reunification was taking hold in Moldavia. The process of reunifying the people who had been separated in 1945 was—at the end of the century—at last possible.

The same dream lived on in the Karelians of Finland, whose country was divided in 1945, at the close of the Soviet-Finnish War, when the USSR appropriated a large part of Karelia's territory. When Germans demanded unification, the Finnish Karelians urged their president to press for the same sort of reunification for them that the USSR had accepted for Germany. A tenth of the population of the Autonomous Republic of Karelia (81,000) is Karelian.[44] Their number decreases every year, and their Finnish compatriots persuasively argued that if they remained in the USSR, they soon would disappear. Although it is unlikely that the Finnish government would demand that Moscow return this lost territory, the Soviet Karelians could easily organize to demand Karelia's sovereignty. Since the number of Karelians was small, their threat to Moscow here was insubstantial. Given the extent of the turmoil throughout the USSR, however, Moscow did not wish to add a new crisis, even a marginal one, to the crises already disturbing the country.

Thus, the facts of the Baltic case argued in favor an initial appeasement that would require time and a distinct willingness to compromise. Whatever the outcome of the negotiations begun between Gorbachev and the Baltic states, their determination to recover the independence they lost in 1940 would change the Soviet Union forever. This crisis signaled the end of federalism, forced the Soviet government to reorganize relations within its former empire, and admit the possibility of separating from it, even if this required following very difficult procedures.

What can be learned from this crisis? First, the central government's behavior and Gorbachev's himself were decisive elements in the crisis. By ignoring what was happening at the periphery—the mounting national aspirations that at first were simple demands for real autonomy—Gorbachev fanned nationalist fires and convinced some republics that no change would come from the center but would have to come from them and in a radical manner.[45] Lithuania's rupture thus was inseparable from Moscow's wait-and-see policy.

Here Gorbachev played a double and contradictory game, and the Lithuanians drew a lesson from it. When he recognized in Vilnius that the Soviet federation was only a myth that had never been established, he seemed to commit himself to granting the republic the real powers that they had never acquired. At the same time, however, while the republics awaited decentralization, Gorbachev had the Congress of the Peoples' Deputies hastily vote for a political reform instituting a presidential power in the USSR. This presidency, which he passionately lobbied for, consolidated considerable central authority in the office of president. The president of the USSR had the power to impose a state of emergency on any part of the Soviet territory—a *presidential law* that temporarily replaced every other authority. It seemed impossible to reconcile real sovereignty for the republics with an all-powerful president who could, if he thought it necessary, arrogate all powers to himself.

The Lithuanians understood this contradiction perfectly. They realized that by creating the presidency before proceeding with any reforms of the federation, Gorbachev was preparing for a "presidential federation" whose contours he could draw as he pleased. This authoritarian idea of the future Soviet system was not opposed just from the periphery. In Russia itself, Boris Yeltsin clearly laid out the problem: creating a presidential power without knowing the territory over which it would apply or the conditions under which the republics would accept being integrated into this space would surely end in the country's disintegration. In fact, the Lithuanians were persuaded that once the presidency was created and able to define the federal framework, it would be harder to leave the federation, so they hurried to leave it before the voting took place on establishing this presidency. In the short term, the Lithuanians' action could be considered extremely rash, but in the long term it was in fact through caution that the Lithuanians staked their all.

The conflict in which they were engaged was surely dangerous, for nothing, other than the spirit of compromise the USSR had shown in Eastern Europe, was reassuring. But would what was acceptable to Moscow outside Soviet borders be acceptable within them? Would the USSR tolerate each republic demanding the right to be treated like Poland and Hungary? The Lithuanians thought they had an effective argument for appeasing Moscow on this point. By referring to the conditions of their 1940 incorporation into the USSR and the 1922 Treaty of Union, they offered Gorbachev a line of defense: what was valid for the Baltic states was not valid for those who in 1922, willingly or reluctantly, jointly founded the Soviet federation. By rejecting this argument and placing Lithuania in the same category as the other republics, Gorbachev was certainly making it possible, for the time being, to challenge its claims to independence; in the long run, however, he authorized all the republics, in case Lithuania managed to make good its exit, to opt for this model and decide on their destiny by themselves. Who was the most incautious in this case, Gorbachev or the Lithuanians?[46]

The problem of caution also arose when the negotiations began.[47] By accepting a temporary moratorium on the effects of independence, was Lithuania pushing its advantage to open negotiations, or was it giving Gorbachev a chance to gain time and defuse independence? No doubt both possibilities existed. But history favored the Lithuanians. Self-determination for Eastern Europe, sovereignty for the republics of the USSR, and all at an ever-speedier pace: it was unimaginable that this movement would suddenly stop and the pendulum swing in the other direction.

At the time of the moratorium, the encounter between the Balts and Moscow primarily revealed their different viewpoints and strategies. The

Balts were momentarily halting their race toward separation, but their achievement was already considerable: independence was part of the political landscape, and the laws that proclaimed this independence had been passed; this moratorium did not affect that achievement. Because they proceeded by taking small steps toward independence and gaining recognition that independence had been achieved, it was of paramount importance to them was that no act or law be "frozen." This moratorium did not impose that on them.

Gorbachev, on the other hand, seemed mainly interested in appearances. It was the explosive charge of words, more than facts, that horrified him. For Moscow, the moratorium meant that the word *independence* was banished, at least temporarily. The laws on the books mattered little. After all, things turned out this way in the test of strength between David and Goliath. As modest as it was, David's slingshot prevailed over Goliath's strength because it marked the victory of the real over appearances.

9

Russia versus the USSR

"THE oppressed will soon stand on the stage of history." At a time when all his work was crumbling, this prediction of Lenin's came true through the sudden entrance, or rather return, of the most unexpected actor to that stage—Russia. In the latest Soviet agony, Russia, subsumed to the Soviet federation since 1917, abruptly became an important part of the national crisis.

At first sight, this return seemed paradoxical. Russia was seen as the "hard core" of the USSR. All the national groups of the union who showed growing hostility toward the Russians and boldly enacted laws limiting the influence of Russian culture and even the use of its language and who urged that the "Russian question" be settled once and for all by disengaging from Russia made Russia the symbol of the oppression they rejected. There was equal incomprehension outside the USSR: although popular rebellion within the USSR seemed a step in the direction of progress, Russia's awakening to this movement immediately stirred up distrust. Russia was disturbing. No one wanted to see this awakening as a chance for it to resume its long, hard journey toward a European destiny.

Salvaging the Past

The reawakening of Russian national feeling is a recent phenomenon[1] that initially expressed only the fears of a small group of intellectuals who rallied around a few themes as they reconsidered the fate of Russia. First, they noted the destruction of buildings and regions bearing witness to their country's past. Churches since 1917 had been systematically demolished or diverted from their original purpose, transformed into warehouses or movie theaters, or simply abandoned and were now crumbling from general neglect. Villages were dying or being transformed by modernization projects. Moreover, in the early 1980s, what remained of the Russian heritage was threatened by the Soviet government's "Promethean" plan to divert the rivers of Siberia to the southern regions of the empire and thus transform those regions into "new Californias." From Stalin to Brezhnev, Moscow had lavishly supported projects intended to show that "socialist man" could transform society and nature as he wished. Each project helped destroy a part of the heritage of the past—architectural and natural. Although no one had cared much until the 1970s, at that point a demoralized society without landmarks began looking to its past to link up with it. Russian intellectuals then discovered how dilapidated their cultural heritage was.

Another reason for pursuing a quest for the past was the dangerous conditions evident in the present. For the Russians, the censuses of 1970 and 1979—so encouraging to the many Soviet groups who had experienced demographic growth and saw a dynamic future—were a terrible experience. The numbers incontestably showed that the Russian nation was shrinking within the empire it had once built and that its population had decreased relative to the populations of the people it dominated. Accused of being domineering, Russians learned that they were in decline. The most observant had a good idea of the causes for this alarming reversal. Russia was exhausted, demoralized, and rife with alcoholism as never before. The family was disrupted, conventional morality no longer existed, and children had few options in a world in which housing was unobtainable, divorce widespread, and day-care centers practically nonexistent. The primary source of this social disintegration was the destruction of villages where families and neighbors grew up in real communities. In the face of this decline, a remarkable group of writers—the ruralists, nearly all originating in the countryside—searched for causes and the dying villages they wished to memorialize. Belov, Astafiev, Rasputin, Soloukhin, and Zalygin were initially, so to speak, archivists searching for a past and a civilization that they thought lost. This nostalgia did not result in a movement or even a program. It was a haunting grievance.

Gorbachev's coming to power and his declaration of the right to the truth changed everything and prompted the dreamers go forward from mourning the lost nation to taking action to save it. They acted on three fronts.

Their most urgent priority was the battle to save the remaining land and villages from the disastrous project of diverting rivers. In this battle, the ruralist writers worked with one of the most remarkable personalities in Russia, the great Byzantine specialist Dimitri Likhachev. Together published an appeal to reason and won their first victory.[2] The project was soon abandoned, despite pressures from Central Asia, where people were counting on the diverted rivers to irrigate their arid soil. Meanwhile, Chernobyl demonstrated how dangerous the USSR could be and opened the way for efforts to save humanity from imperfectly manipulated progress.

The second priority was to recapture the past. For years history—both recent and distant—had been concealed. Noting that Russia had lost its peasant roots—as well as its roots, period—the ruralists were not only going to reclaim Russian history but return it to society. Worse than the destruction of the villages was a history that was primarily a veritable genocide of Russian peasantry. If genocide means the systematic destruction of an ethnic group, then the term fully applies to what was suffered by the Russian and Ukrainian peasantry, which was liquidated precisely because it was peasant and because Russia's foundation was its peasant civilization:

> Nearly all the capital of intellectual energy accumulated in Russia during the nineteenth century and used for the revolution was frittered away on the peasant masses. The intellectual who produces spiritual sustenance and the worker who creates the mechanisms of the urban culture are increasingly being devoured by the peasantry who voraciously feed on what others have produced through incredible effort. It can be said with certainty that the peasantry revived itself by killing the intelligentsia and the working class. . . . The Russian people of the cities and villages, half-savage beasts, stupid, almost frightening, will die to make room for a new human race.[3]

These hateful, racist words of Maxim Gorky, a friend of Lenin, concerning the peasantry were published in 1922 and presaged the policy to come. The fight waged by the ruralist writers led them eventually to publish remarkable books describing this genocide, exposing those who were responsible (Soloukhin put Lenin on trial on this charge), and evoking the current abandonment of the countryside. They also are credited with revealing the publication of the most abhorrent writings, such as those of Gorky, hitherto concealed from criticism but now inescapably linked to the genocide of the peasants.

The ruralists succeeded in showing that the country's unbridled industrialization cost the deaths of tens of millions, the pillaging and degradation of peasants into slavery, and ultimately the destruction of the roots of the Russian civilization.[4]

This leads to the ruralists' third issue, where they focused all effort on a single question: Why did this happen, and did Russia deserve it?

The answer lies in the real history of the country, the one that was hidden away in 1917. Once again, the great intellectual Dimitri Likhachev, a model of tolerance, found himself in the company of people whose nationalism verged on chauvinism, to demand the immediate publication of Karamzin's *History of the Russian State.*[5] Then the people could determine the meaning of their past and whether the Russian state offered a progressive alternative to the radical solutions of Lenin.

These first struggles gradually gave rise to Russian national feeling but not a mobilization against other peoples. Their primary purpose was to revive and comprehend the past. Certainly, though, an issue of future conflict was becoming apparent. Did this ravaged past contain the seeds of progress? If so, by what route—the distinctive way of Russia or the Western way, which evidently had led to disaster. The old conflict about being for or against Westernization was looming.

Aside from a few aberrations, however, the Western influence had not yet been equated with a "Judeo-Masonic conspiracy." When Dimitri Likhachev said that "nationalism must actually be patriotism"—that is, loyalty to the nation but not isolation from others—he was fairly summarizing the spirit that animated this quest for the truth of Russia. Intolerance would come later.

Looking Russia in the Face

Uncovering the tragedy that Russia had suffered was a terrible ordeal. It would have been virtually impossible to remain calm in the face of revelations about the systematic massacre of Russians, the contempt and degradation suffered by an entire society, and that society's moral complicity with its executioners.

Because the government was Russian and all the Soviet people considered that Russia and the Soviet system were two aspects of the same thing, the Russian people could not avoid acknowledging that they had been both victim and executioner. After decades of misery, Russia had to learn about itself, calibrate the extent of its misery, and accept its responsibility so that it could eventually accept itself. Through its reawakening, Russia had to arrive at the essential truth. This gave its nationalism a special tone: it was not the triumphalist nationalism of a people that have had a revelation of their power, but a nationalism of suffering and shame.

As if that were not enough, Russia's rebirth is set in a framework of disaster that ultimately may doom the nation. Not only is the population caught in a declining trend that in the short term no magic wand can reverse, but the Russians are also a physically and intellectually "damaged" people.

Russia shared with Lithuania, Estonia, and the Ukraine the grim distinction of having the highest mortality rates in the Soviet Union. They are much higher the national average (10.5 for every 1,000 people in Russia versus 7 out of 1,000 in Central Asia in 1987), which is unsurprising when the age pyramid is taken into account. Rates for Russian alcoholism were also the highest, and the number of alcoholics even increased during the years of perestroika.[6]

Although this decline in health was foreseeable because of the high degree of urbanization among Russians, their intellectual decline was surprising. A recent book gives a clearer idea of the situation:

> The census of 1970 showed that while, among people over sixty, the Russian republic had the highest percentage of college graduates in the USSR, the situation was reversed in the younger age groups and, for the twenty to twenty-nine year olds, Russia was behind all the republics except for the Ukraine and Kirghizia. The 1979 census strongly highlighted this trend. . . . The percentage of specialists with a secondary

education or higher is above the national average for the Caucasus and Central Asia, and below it for Russia.[7]

All educational indicators are in the danger zone. On the average, Russia had half the number of institutions of higher learning than each of the union's other republics. It is poorer in doctoral candidates and Ph.D.'s and was fifth in the number of members and correspondents of the Academy of Sciences.[8] Primary and secondary education are also neglected; although the number of children in school is growing very fast in Central Asia, in Russia it has fallen off. The compulsory ten years of schooling have not checked this "de-education," for that is what the phenomenon is. This involves the rise in the number of children with physical or mental disabilities, the closing of schools in sparsely populated areas, and the low level or absence of the teaching staff in the less prestigious schools. On the periphery, on the other hand, the number of schools is increasing and the teachers' proficiency is improving.

The effects of this trend are a decline in the Russian standard of living and an increase in that of the peripheral societies: 40 percent of the Russian working population is employed in industry, as compared with 10 percent or less in Central Asia. In the southern periphery, the working population is divided between intellectuals and peasants. While the working class stagnates, intellectuals enjoy great opportunities for social advancement. The peasants represent the moral stability and the permanent social values that give society security. In other words, to be Russian offers few—or fewer—chances to take someday a powerful place in the federation.[9] Thus, it is not just in demography that the Russian people are losing their leading role in the Soviet Union; there is also a regression based on real differences.[10]

Long accused by the nationalities of dominating the union, the Russians were unable to gauge the accuracy of this observation. Glasnost, however, made it possible to confirm the intuitions of the national intelligentsia, who were convinced that Russia obtained no benefit from the federal system. Gorbachev had already pointed out that the "parasitism" of the republics was intolerable. To the Russians, however, this was less serious than the intellectual destitution of which they knew they were victims.

This humiliating status had been created by the very strategy of federalism. For one thing, the advancement of other ethnic groups in the USSR, as in the United States, had effects that were far too radical. In some of

Moscow's more prestigious institutions of higher learning, the inegalitarian admissions policy favoring other nationalities sometimes meant that no Russians were accepted.[11] In addition to advancing people from other nations was the desire to identify Russia with the USSR so that the same opportunities enjoyed by Russians could be made available to all the nations, a desire that deprived Russia of many institutions that existed elsewhere. Until 1990 it did not have its own Communist party or Academy of Sciences. In the republics, the academies of sciences are devoted to the advancement of the nation. Each federated republic thus had an institute of history that recorded the nation's history in great detail. This inequality led to situations that were, to say the least, startling. Soviet libraries were filled with books about World War II, such as *The Contribution of the Karakalpaks to the Second World War* and *The Uzbeks in the Second World War*. No people, however small, lacked national institutes that extolled their glory and exploits, except the Russians. To judge by the lack of books in the libraries, Russia had done nothing in the revolution, the war, or the economy. The Russian people as such did not exist.

The fiction of its identification with the USSR not only stripped these people of their history but also damaged their culture and language. No institution existed for the advancement of Russian literature, which was the common property of the Soviet people, who, as we have noted, did not exist. As a result, the Russian language, imposed on all as a lingua franca, has become extremely disfigured. As the peasantry was thrown into the cities, subject to an unprecedented intermixing of populations, and forced to go into exile beyond its borders, the Russian language has been impoverished and "peasantized." To made it comprehensible to all citizens of the USSR, Russian was cluttered with abbreviations, contractions, and Sovietisms. While the border peoples battled to save their languages from foreign imports and maintain their grammatical and syntactic purity, Russian was transformed into a veritable gibberish scornfully rejected by the nationalities. Again, no Academy of Sciences preserved its heritage.[12]

When, at the end of this century, Russians close the books on their domination of the federation, they will observe that their material, cultural, and intellectual poverty are the results of the horror they have experienced. They have to accept all this. It is the framework within which a national rebirth will take shape.

Gorbachev and Russian Nationalism

Mikhail Gorbachev could not remain indifferent to the rise in Russian anxiety that coincided with his coming to power. He decided to use the prestige of the intellectuals engaged in the Russian battle to help convince the people that their efforts were needed. This prestige was measurable by the readership of their books. In the USSR people fought about the books of Pikul Astafiev, Belov, or the Kirghizian writer Chingis Aitmatov, much beloved by the Russian writers. Books about the reality of their peoples have the highest sales in the USSR—ten times greater than Gorbachev's *Perestroika.* These writings, which fascinate society, cannot be ignored.

This official acknowledgment of Russian national feeling had two immediate consequences: it appealed to certain of its representatives and it promised a reconciliation with the Russian Orthodox Church.

In the fall of 1986, Gorbachev named the academician Likhachev, considered the "conscience of Russia," as president of the Fund for Soviet Culture, and the ruralist writer Sergei Zalygin became editor-in-chief of *Novy mir.* These two nominations should in principle have won the support of the Russian intelligentsia for Gorbachev.

Under Zalygin's leadership, *Novy mir* was really the journal of Russia's moral and intellectual rebirth. All the trends that explain the past were expressed in this journal. It published the prohibited books of Boris Pasternak, Bulgakov, and Solzhenitsyn. The journal assigned Sergei Averintsev, another emblematic figure of Russian liberal thought, with the responsibility for rediscovering and publishing the texts of nineteenth-century thinkers.[13]

Once Russia linked up with its heritage, it could draw on it, rediscover itself, and discover a new direction. The Fund supervised by Likhachev also offered opportunities for national projects. He intended to revivify the past by restoring the old names of streets and cities. The rehabilitation of the great poet Gumilev, who was shot under Lenin, and the publication of the works of Karamzin all suddenly seemed to be possible because respected men directed the journals or institutions that could promote this policy for rediscovering Russia.

Gorbachev's efforts also involved a reconciliation with the church. Since the "moral concordat" signed by Stalin and the Russian Orthodox church, relations between these ill-matched partners had fluctuated. The

Russian Church had regained its right to exist but at a shameful price. It was a patriarchate subject to the state, unrespected and ever ready to associate itself with any governmental action, such as appeals for peace, and so forth. Its clergy had notorious links to the KGB. For years only the elderly faithful attended services at its modest number of churches. Everywhere, the government perpetually vacillated between desiring to eliminate religion—Khrushchev really attempted this—and showing a tolerance that compromised the hierarchy.

Nevertheless, in the early 1980s, the grim state of the church included a few bright spots. The Russian intelligentsia hesitated less and less about returning to its faith. Concerned young people overran the church. They often came there by a circuitous route, after discovering that the decaying church structures needed salvaging. Gorbachev saw a twofold attraction in the Russian church. As a Russian, it was a way to present himself as a defender of his country's religious roots. Because the church offered a system of moral values—the time had not yet come to speak of spiritual values—it could help him invigorate a disheartened society. That explains why the millennium of the Christianization of Russia was for him an opportunity to create a bridge between the church and the Soviet government. He lent his support to a celebration of magnificent pomp, closer to the splendors of the czars than those of the USSR. His formal meeting with the patriarch and the Holy Synod,[14] the presence of Raisa Gorbachev at the ceremonies at the Bolshoi Theater on June 10, 1988, Gorbachev's systematic criticism of the curbs on religious freedom, and the news, given to the French press, that he had been baptized all helped persuade the church to support perestroika, now regarded as one of the great national institutions. The healing of Russia, the restoration to the church of its long-closed houses of worship, the authorization to build new churches and some day to celebrate—unforgettable to Russian hearts—a service in the Kremlin's Assumption Cathedral were all signs to the Russian people that it was finally blessed with a government that understood it.

This probably was not enough to persuade the Russian intelligentsia to back perestroika. Even though it considered Likhachev its representative at the Fund for Soviet Culture and *Novy Mir* the most appropriate vehicle for expressing its ideas, it insisted on founding institutions that were explicitly Russian and without the word *soviet.* Thus were born the Association of Russian Artists and the Foundation for Slavic Literature and Culture.

The Association was founded in November 1988 under the guidance of a group of well-known Russian intellectuals—Rasputin, Belov, Astafiev, Kuniaev, Lobanov, Kozhinov. The editor of the very conservative journal *Molodaia Gvardia* was to direct its efforts at rediscovering and disseminating Russia's cultural and historical patrimony and mobilizing the intellectuals in safeguarding the country's unity from nationalistic pressures. The Foundation was born of cooperation between several institutions—the Writers' Unions of the three Slavic republics, the Russian Orthodox Church, the Church of the Old Believers—and wanted to join the effort to heal the Slavic peoples of Russia. At its head were three writers who were associates of Gorbachev—Rasputin, the Ukrainian Boris Olinik, and the Belorussian Nil Gilevich. It had the explicit support of Alexis, the patriachate of Moscow who in 1990 succeeded the patriarch Pimen, which underscored the blend of religious and pan-Slavic ideas that characterized the undertaking.

Gorbachev did not so much win over the promoters of the Russian cause; rather he helped them, by his opening in their direction, to gain control of important media outlets and, in the final analysis, to give legitimacy to Russian nationalism.

Russian National Pluralism

The West is mainly acquainted with the extreme and sometimes caricaturish type of Russian nationalism identified with Pamyat—a kind of organized, intellectually coherent neo-Nazi movement. Pamyat is, in fact, a name for a myriad of distinct organizations, which we shall return to later, but the most influential brands of nationalism are all found outside Pamyat and are considerably different from the stereotypes commonly held in the West.

Russian nationalism, as it developed from 1986 to 1990, was roughly divided into four main branches—liberal nationalism, radical nationalism of both the right and the left, and conservative nationalism. All share certain demands or projects that presently constitute the program of Russian nationalism:

- An independent Russian government with local governments and broad economic autonomy in Russia for the expression of individual initiative;

- The restoration of land ownership to peasants;
- Genuine separation of church and state;
- A return to traditional Russian names for towns and streets.[15]

In addition to this shared program, each type of nationalism had its own views on problems as important as the choice between the Western model of development or the Russian socioeconomic tradition, the place of religion in the political system, and relations with other nations.

The moral leader of the liberals, who could be described as Christian democrats, is Dimitri Likhachev, who, beyond the struggle to preserve the cultural and natural heritage, attempted to keep the various nationalisms from indulging in chauvinistic leanings. Aided by Zalygin and the staff of *Novy Mir,* he has sought to keep in the ranks of liberal nationalism the people he thinks could attract many sympathizers through their talent and comprehension of the Russian tragedy. Rasputin (despite some excessive language), Belov, and Astafiev remain close to this group because Likhachev does not wish to push them into the welcoming ranks of the conservatives. For this branch of nationalism, the early twentieth-century intellectual and philosophical Russian heritage is sufficiently rich to serve as a reference point for a Russia in search of itself. Berdiaev, Father Bulgakov, and Semion Franck were three philosophers who were able to find an alternative to Marxism in Christian values. Unknown in the USSR, they have enhanced renascent Russian thought. Likhachev has said that this Christianity is the key to the Russian political culture; it has no place for intolerance and the rejection of others. Quite the contrary, it contains the elements for bringing together people whose history has been tragic. The liberals are patriots—whose homeland is not defined by either its territory or its sufferings but by an identification with a cultural heritage—and vigorously disown the whole of Marxism, particularly its "annihilation of nationhood."[16] But they also reject the chauvinism that may lead to anti-Semitism. Likhachev said that this last attribute was peculiar to Russian "pseudointellectuals" rather than the true intelligentsia, which, like the true peasantry, is free of it. The openness of Christianity does not condone such things. Hence the need to rely on the authentic heritage of Russian culture. Gorbachev himself said in a speech that "Russia was the last refuge of spirituality."[17]

At the other end of the political spectrum, the radical nationalists or nationalist Bolsheviks are not Stalinist, but they do not reject the heritage of their former communist leaders. They credit these leaders with having maintained the empire and building a solid state that retained many of the traditions of the preceding state. Certainly, they reject Marxism-Leninism, but they implicitly accept a Russian variant of it. They occupy key posts in the major cultural institutions—the Writers' Union of the USSR and the Writers' Union of Russia. They put out a newspaper, the *Molodaia Gvardia,* and are often featured in the army newspaper *Krasnaia Zvezda.* Their leaders, Bondarev and Poskurin, attempt—as do the liberals—to woo the hesitant center, which is too facilely called rightist.

Also radical, but located on the left, is a group that meets in clubs that proliferated in Moscow and St. Petersburg and that see themselves as reformers of the social-democrat and Westernizing type. In publications like *Mozkovskie Novosti, Oktiabr,* and *Znamia* Gorbachev found his warmest supporters. They attempt to determine the borders between the achievements of the revolution and the "cult of personality." In a certain way, the *Children of the Arbat,* Rybakov's saga, reflects their program. What unites them with Gorbachev is the attempt to shed light on the conditions of Stalinism and the destruction of the Leninist old guard and to glean from these disasters a modernization plan. They struggled to discover figures from the Soviet past—Bukharin, Trotsky—who could give Russian politics a different image from the one of the prorevolutionary Slavophiles.[18] Many of them entered the electoral contest for the Congress of the People's Deputies, where they tried to forge a democratic program with which the Russian people could identify. Their main difference from the liberals is that they see Christianity as a private matter, not a constitutive element of the politics by which Russia will rebuild itself. Another difference is that although they call themselves Russian, nationalism counts less than their attachment to the old Soviet desire for modernization. They wish to salvage something from socialism—such as the desire for unity, Bukharin's ideas, and the NEP [New Economic Policy]—for Russia's development.

Among these different nationalist strategies within Russia there is a center that can be called conservative made up of writers respected for their attempts to restore its past to Russia. This group has attracted the ruralists, the critic Kozhinov, and the reclusive painter Glazunov. They completely reject Marxism-Leninism and everything else that they feel has produced a system of thought that diverted Russia from its destiny. They reject the West that produced Marx and his heirs and that recently has perverted Russia in other ways—through a rootless, overpermissive mass culture, cosmopolitan rock, and drugs. They see the moral values of

Christianity, so enrooted in Russia, as a refuge against a pseudoculture that has no goal other than to entertain the human spirit and divert it from its true purposes—personal salvation and the preservation of a culturally stable community.

Like all the nationalist trends, this strategy—whose hopes for the future are based on the traditional peasant and Christian culture—devotes an enormous amount of time to analyzing the past. Here is where conflicts and accusations appear. These fervent nationalists fight against the cult of personality by seeking the deeper reasons for this cult. This thinking links them to Solzhenitsyn, and they insist on the existence of a nihilist plan for the destruction of Russian society.

The critic Vadim Kozhinov considers this problem in his brilliant essay, "Pravda-Istina."[19] His analysis has produced a tempestuous response and set the stage for a kind of war of religions. Accused of anti-Semitism or of being a fanatic from the "Masonic conspiracy," Kozhinov's language is surely excessive, but he summarizes a very important debate: Was the Russian tragedy Stalin's handiwork, or are its origins located in a *system of thought* that led to the revolution? Is Russia, with its backwardness and a people unaccustomed to democracy, responsible for Stalinism? Did Russia offer an ideal breeding ground for it? Or did it involve a deformation, a spiritual perversion alien to Russia and imported into Russia from abroad?

Kozhinov ignores the group of writers who have examined the cult of personality—Rybakov, Mikhail Shatrov *(Onward, Onward, Onward!, The Brest Peace)*. He is embroiled in a controversy with the critic Benedikt Sarnov and is generally opposed to all those who feel that the heirs of Lenin who corrupted his message caused the tragedy. Because of his opposition to writers whose ideas are highly popular among the intelligentsia and because his criticism is directed toward writers (Shatrov, Razgon) and Lenin's associates, many of whom were Jewish, Kozhikov has been accused of anti-Semitism and cast into the Pamyat camp. For the liberals, however, his comments echo the apprehensions of the Russian people and must not be dismissed as extremist. When Rasputin rages against the fact that Pamyat is not only the object of vehement criticisms but of calls to silence it—adding, "All our glasnost does not leave room for a word of defense"—he too is accused of anti-Semitic extremism by the *Moskovskie Novosti*. But he has warm defenders among the liberals. Alla Latynina,[20] who has written perhaps the best synthesis of these conflicts between the intellectuals and who profoundly disagrees with the extreme statements of Kozhinov and Rasputin, nevertheless stresses that the important thing is to confront the whole truth—not to be content to determine the conveniently guilty parties but to delve deeply into ideas. Rasputin has the virtue of "taking on the hard task of asking all the questions, brutally but frankly."

While she is at it, she has a warning for the radical nationalists: "It seems that those who have risen up against the conservatives, waving the flag of democratization, who have toppled the Stalinist edifice, who want an open society, must bet on liberalism. But freedom of thought is not a commodity that can be bought and sold. . . . The theory then is: 'Let us win first and later we shall make a place for freedom.' "[21] In this, she echoes the apprehension expressed by another liberal, A. Strelianyi: "What I most fear is that the conservatives will be silenced. If we get to that point, then we will not be content to imitate them but will become them."[22]

This is very much the position of the liberals, who try to avoid participating in or legitimating any extremism and try to avoid any attempt to purge dissidents who can influence the people.

It is clear that national-liberal thought accepts neither the extremism of Pamyat nor the neo-Stalinism of Nina Andreyeva. In fact, she is not an intellectual but an obscure "educator" and quite unknown before the publication of her manifesto—a tangled condemnation of Gorbachev's reformism which she called "leftist liberalism," of national liberalism, and of all positions that reject the heritage of revolution.[23] This nostalgia for Stalinism, which calls for a speedy return to order, also displays an anti-Semitism that suggests that the Gorbachevian liberal left is dominated by Jews, just like the group of Stalin's opponents.

As for Pamyat,[24] it is a cluster of organizations born in the late 1970s, originally for the purpose of rescuing historical and cultural monuments. For some years, it was content with this goal. Then, under the influence of the photographer Vasilev, a remarkable orator capable of electrifying an audience with speeches that were anti-Semitic and hostile to all the nationalities of the USSR, Pamyat succeeded in mobilizing a large public for a time; but its influence quickly declined as Russian nationalism appeared on the political scene. Pamyat has scarcely any answers for the many people in Russia who feel nostalgia for a lost past, which under present conditions takes on the look of a Golden Age and makes nationalism attractive. Quite the contrary, the entry of the Russian nation into politics, with respectable spokespersons and deputies elected by universal suffrage, diverts attention from the clamorous but ineffective small groups. Pamyat's tendency to split into an ever-increasing number of organizations (*Otechestvo, Rodina, Patriot,* and so on) weakens these groups and their cause. The idea that Pamyat was a tool of the KGB also has been discussed.

That does not mean that the ideas expressed by the supporters of Pamyat, chief among them anti-Semitism, are completely doomed. They also are echoed in the bitterness of the Russian people. Nevertheless, the growing appeal of Christianity provides grist for the mill of a nationalism that is surely conservative but that liberals are trying not to abandon to its extremists. Preventing this drift depends largely on those with a political mandate or a forum. And everyone, whether a sage like Dimitri Likhachev or writers of books and articles, is now calling for a humanistic Russia.

The debate remains unsettled on one central point—whether to be Westernized, fated to imitate Europe. Liberals like Tsipko, who is attempting to eradicate Marxism from the Russian world view, and Selunin, Averintsev, and Latynina defend the restoration of a Christian culture as the foundation of the future Russian society and at the same time reject the idea that Russians should turn their backs on Europe and the West.[25] Even if the idea of a solitary Russia, turned in on itself and its own genius, is attractive, Tsipko's ideas—a combination of a Russian and Christian culture with Western contributions—could take root as Russia's political future takes shape. This gradually pieced-together future will make it possible to halt nationalism's drift toward the anti-Semitic right. Similarly, the temptation of a drift toward neo-Stalinism will also diminish as new Russian leaders find some remedies in their republic, however limited, to disorder and scarcity[26] and as they provide ideological guideposts, as they have begun to do.

The Political Future of the Russian Nation

In the sphere of government, everyone has tried to benefit from this renascent Russia—all the more so as the crises in the nation-republics at the periphery made Russia appear the one certain pillar of the empire, but a pillar that no longer provided support, as in the past, without a price. This price was the recognition of the rights of Russia and of the USSR's debt to it. Gorbachev, the first to grasp its importance, not only surrounded himself with liberals, to whom he gave media access, but, on reaching the presidency, integrated into his Presidential Council some fairly extreme representatives of the Russian nationalist movement. The presence of the writer Rasputin in the council was surely a way to recognize the need to include a spokesperson for Slavophilic nationalism in the majority.[27] Similarly, the presence at Rasputin's side of Chingiz Aitmatov, the Khirgizian writer who was most popular with the Russians and who was close to all Russian national movements, also indicated the growing importance of national positions.

Like the Russian nationalists, liberals, and conservatives, Aitmatov argued for the restoration of religious values and was concerned about the country's moral erosion. But it was difficult to see how the reformist economist Shatalin and the conservative Rasputin would work side by side, with both of them haunted by the idea that any reform that orients the country toward capitalism is dangerous for Russia and would lead it immediately to lose its identity.

Rasputin worked on the liaison between Gorbachev's government and the Orthodox Church. Unlike the other members of the council, he was not a member of the Communist party. He represented the Russia that did not succumb to the temptations of communism that entered the core of the state apparatus.

Another national "nomination" was that of Veniamin Iarin, one of the founders of the Russian Workers' Front created in 1989[28] and representative of the ultraconservative wing of the working class that Gorbachev dissuaded from the temptation to recreate a neo-Stalinist front.

Another gesture toward nationalism was the election of the new Russian Orthodox patriarch, Alexis, as archbishop of Leningrad and Novgorod. It is unlikely that Gorbachev arranged for his election, but the government must have played a decisive role in the naming of a patriarch, for no patriarch could be elected against the government's wishes. Alexis, Estonian in origin, is heedful of national problems and a supporter of the Russian cultural undertakings—one of the new people Gorbachev needed to avoid having nationalism turn against him.[29]

Yegor Ligachev, who since 1985 has assiduously played the role of the conservative trying to uphold the accomplishments of the Soviet system, has been the spokesperson for an unchanged ideology and has opposed the temptations of a religious renewal: "Some people debate the utility of greater religious tolerance. They forget the fundamental teachings of Marxism, which is that in no case is religion the source of morality for the individual."[30] Ligachev's ties to the Bolshevik-nationalists—and even beyond that, with the neo-Stalinists—appear to be well established. Even though no tangible proof has been given, Nina Andreyeva clearly recognizes his authority,[31] which puts him in the uncomfortable position of seeming to be the chief figure of a neo-Stalinist movement rather than a critic of nationalism. At a time when nationalism appears to be the surest way to establish a favorable reputation with the people, this orientation is unfortunate, to say the least. John Dunlop, a leading expert on Russian nationalism, has revealed, however, that through an official visit to the Glazunov exhibition in July 1988, Ligachev perhaps found the way to join the camp of the "hard-line" nationalists.[32]

But the USSR's political development, notably the public debate over its elections, was to shift policy making from the intelligentsia to the people.

Public Opinion Revealed: The Elections of 1990

The local elections of 1990 were significant in Russia because the republic now had institutions that duplicated those of the USSR. Russia had a bicameral legislature—the Congress of the People's Deputies, made of 1,068 elected representatives, and the Supreme Soviet, with 152 members.[33] Two major opposing camps confronted each other during the campaign—the Bloc for a Democratic Russia and the Bloc of Russian Patriots.

In Moscow this face-off was particularly interesting and significant. The Bloc for a Democratic Russia included among its Muscovite candidates some leading liberal figures, including the editors of several publications whose success was due to their openmindedness. The editor-in-chief of *Argumenty i Fakty* (33 million subscribers), Starkov, some of its editorial board, and the editors-in-chief of *Vek XX i Mir, Sel'skaia Molodezh,* and *Voprosy Ekonomiki* offered themselves as candidates. The candidates also included many activists from the Interregional Group of the Soviet Congress, which is at the forefront of reformism. The Russian election mobilized everything that the Russian politicians considered most prestigious in the liberal camp.

The Bloc for a Democratic Russia, formed in January 1990 to prepare for the elections, adopted a program oriented chiefly toward democratic reforms—multipartyism, decentralization, an end to economic and political monopoly. But a nationalist Russian party also was present: "We shall call for true sovereignty for Russia with the primacy of the republic's law over federal law."[34] On the whole, under the heading of nationalist demands, the Bloc for a Democratic Russia adopted all the demands of national liberalism, but in a moderate tone in which the desire not to clash with any people, and the desire to avoid chauvinism, prevailed.

The Bloc of Russian Patriots was an alliance of cultural and political associations formed at the time of these elections. Despite the word *Patriots,* to which the academician Dimitri Likhachev gave an open and tolerant meaning, this alliance was characterized by closure and intolerance. The Russian Patriots—their electoral platform was explicit about this—criticized the economic reforms that resembled the free market, while defending the idea of peasant land ownership; they opposed the weakening and dismantling of the Soviet state and empire. Finally, they demanded that Russia play a central role in the Soviet state. Inconsistent among themselves, however, they also asserted that Russia needed an independent status. They valued conventional morality, the reinstatement of spiritual values, and a return to Russian cultural traditions. This platform contradictorily combined the aspirations of the Bolshevik nationalists and a certain traditional Russian populism,[35] possibly in the hope of attracting two contrary voting blocs—people nostalgic for Stalinism like Nina Andreyeva and those sympathizing with conservative nationalism. This muddled calculation no doubt explains the electoral disaster suffered by the Russian Patriots and the victory of the Democratic bloc.

This defeat of the hard-line nationalists was all the more remarkable because they ran the loudest campaign and also received the most media support. Throughout the entire campaign, the Party organ in Russia, *Sovetskaia Rossiia* threw its weight behind the Patriots. Their defeat was a disaster for the Party, for it showed the Russian population's dislike for the variant of nationalism the Party backed. Nearly all the stars of the conservative side were beaten; a small number won in distant voting areas. To get elected, the editor-in-chief of the *Sovetskaia Rossiia* had to run from an obscure district of Daghestan, where probably no one had ever heard of him. Writers like Bondarenko, Salutsky, and Glushkova were roundly defeated (winning from 2.72 to 3.41 percent of the votes).

The electorate was no more indulgent with the Pamyat candidate, who received 5 percent of the votes. Nevertheless, this debacle yielded a lesson. The candidates who emphasized their attachment to Russia and not nostalgia for the lost order—such as Kuniaev, editor-in-chief of *Nash*

Sovremennik or the painter Glazunov—obtained the most votes, even if they did not manage to get themselves elected.

At the first round, the Democratic bloc had winning candidates who were known for their ultraliberalism, like the young economist Tatiana Koriagina, who is one of the bright stars of Russian politics (65 percent of the votes on the first round). The voters' preferences were clear. This balloting was noteworthy for the very great differences in votes between the winners and the losers. Against several opponents, Starkov, editor-in-chief of *Argumenty i Fakty,* was elected with close to 51 percent of the votes. On the other hand, the highest number of votes for the defeated candidates from the Bloc of Russian Patriots was about 20 percent; only one of its candidates reached 41 percent.

The elections produced a large majority for the Bloc for a Democratic Russia in Leningrad (in the two largest cities of Russia, the Democrats also won the municipal elections) and numerous seats throughout the rest of the country. Thus, a fairly open congress was selected by the voting.[36]

But the voters' preferences were less simple than they initially seemed. They were no more attracted to liberals than conservatives but rather to candidates who could combine an open discourse with the demand for a nationalist Russia. The Democratic bloc could measure the importance of this demand and its need to occupy the national arena at the elections to the Soviet of the Republic and especially the naming of its president. These elections were indeed disappointing for the Democrats, whose leading candidates were beaten and who found themselves a minority in an assembly dominated by conservatives. This defeat made it all the more important to win the presidency, which—as Gorbachev had shown in the Supreme Soviet of the USSR—could, if handled adroitly, impose its views on any majority. The election was decisive because Russia as yet had no "head of state," so its "strong man" was the president of the Supreme Soviet. A defeat here seemed unavoidable in the face of a conservative majority.

At the first session of the congress on May 16, the liberals won a decisive victory when the assembly aggreed to an assessment of the record of the outgoing government. This government, led by the ultra-conservative presidential candidate Aleksandr Vlasov, would not fare well under such an assessment. Its disastrous policies condemned it, discrediting Vlasov's candidacy and leaving the way open to Boris Yeltsin, whose candidacy had finally won the democrats' approval. Thus, the battle for the presidency was between two candidates, Yeltsin and Vlasov,[37] who both—and this is what is important—placed Russia's sovereignty among their top priorities. But whereas Yeltsin said, "total sovereignty," Vlasov replied, "economic sovereignty within the framework of the Soviet political system." Strongly tinged with nationalism, Yeltsin's speechmaking won him 537 votes on the first round, as against 467 for his opponent.[38]

Yeltsin's first speech confirmed his Russian game plan: a hundred days to establish the republic's sovereignty; the introduction of a presidential system in Russia that made the republic's president as powerful as the Soviet president; very swift reforms, so that Russia moved ahead of the USSR in its reconstruction efforts; restoration of the moral authority of the Russian Church,[39] notably by giving it a place in the educational system;[40] and criticism of Gorbachev's too-pronounced orientation to the West.

Yeltsin's speech encouraged the growing Russian nationalism. And Yeltsin insisted that he was the president of all the Russias, all political factions together, and not only of the most reformist party. We can thus compress this speech into a single popularity-seeking desire to strengthen an already solid position. By using language that was more nationalist than reform-oriented—even though Yeltsin also wanted to present himself as one who would push for ever greater changes—the new Russian president was aware that he was going in the direction in which all Russia, both liberal and conservative, was headed.[41]

The growing number of extremist organizations (despite Pamyat's electoral defeat) may have convinced the Russian liberals that if they relegated the problem of Russian nationalism too much to second rank, they might someday be overwhelmed by demonstrations of radical chauvinism. It appeared urgent to bar the way as quickly as possible to a nationalism that would be unacceptable because of its excesses and thereby to prevent a de facto alliance from forming between this extreme nationalism and the conservatives. Indeed, after the elections, the liberals realized that their victory would prompt the conservatives to play the nationalist card and find fertile ground for winning back public opinion. The relative democratization of the USSR that allowed not only electoral competition but also the development of parties and political expression[42] had the highly unexpected effect of opening the floodgates to a hitherto amorphous nationalism. As scattered polling showed, public opinion (outside the problems of daily life) was primarily interested in Russia, its rebirth, and its self-expression, and politicians of all persuasions were condemned to respond to this problem or be cut off from the people. The liberals, who were somewhat behind in this matter, had to work out a genuinely Russian platform. Yeltsin took this path early on, and his political position in Russia—insofar as the economic successes take time—will depend primarily on his ability to establish himself as the nation's defender and win acceptance of his ideas about Russia. The governments of the Baltic states realized this and warmly greeted Russia's independent orientation, convinced that it would strengthen their own cause by weakening the federation.

Suddenly, the liberals found themselves in a difficult position. They had to play the Russian card, but out of allegiance to their own ideas, they also had to try not to stir up a narrow nationalism that could turn into hostility against the other peoples of the USSR. It would be quite a feat to keep Russians away from the temptations of chauvinism and isolationism without committing any of these errors. Many intellectuals were alarmed. Their country was not a little Estonia but a powerful state with no need to defend itself against any other. By confusing its problems with those of the smaller nations of the USSR, it risked lowering itself to their status.

It was, in fact, Aleksandr Tsipko—a liberal nationalist and Christian who had lucidly analyzed the Marxist roots of the Soviet tragedy—who came to the most stimulating conclusion in this debate, in his article "The Russians Would Leave Russia":[43]

> The RSFSR's leaving the USSR would not resolve any Russian problem; Russia's sovereignty solves nothing. If the roots of Russia's unhappiness are not eradicated, the independent Russians will always be as unhappy as they are today, integrated into the federation. Furthermore, Russia is not Lithuania. Vilnius can dissociate itself from Moscow, and that is its right. But Moscow cannot leave Moscow. Moscow can only go back to its history and free itself of what has so long prevented it from living and growing.

The rejection of the federal system that began with the emergence of a second government—the one the popular fronts were gradually implementing—thus ended in two kinds of rupture: the independence of Lithuania and then that of the two other Baltic states, and the proclamation of Russian sovereignty.

These two ruptures, of course, do not have the same significance. In proclaiming their independence, the Baltic countries really intended to leave what remained of the USSR. They did not have any real affinity with it, and the ties that for fifty years bound them to the federation had been imposed by force, although they had achieved a proper balance in 1918. Their culture and interests oriented them toward Scandinavia. Because postwar history forced them to integrate with the USSR, they were so economically interdependent with it that a total breakaway would be unreasonable. Knowing this, the Balts tried to leave the Soviet Union by invoking their distinctiveness and so avoiding a divorce in the confrontation.[44] They were convinced that, after an initial crisis, arrangements could be negotiated on trade, port facilities, and even the status of the Russians who chose to remain, and that the old ties of dependence would be replaced by the contractual ties of cooperation and good neighborship. As the experiences of the British Commonwealth and the Community of French-speaking States have shown, once the time of domination is over, people who have shared the same fate can maintain certain aspects of the old life. When the European landscape, believed to be fixed forever, changed, the Baltics, incorporated within the USSR just when this landscape had been fixed, should be able to participate in this change. If German unity, which seemed permanently a thing of the past, was now acceptable to Moscow, why not Baltic independence? Seeing the limits of the impossible shrink each day in Eastern Europe, the Balts judged, reasonably enough, that the limits should shrink for them as well. This reasoning was all the more valid because the glue of federalism—the political system, the law, the army—was gradually disappearing. A twofold government and pluralism had insidiously overrun the political space and replaced the Soviet system.

Officially, this system was intact, but in reality, it was just an empty shell. Even before independence, elements of another system had slipped in that had nearly nothing in common with the USSR. No one now obeyed Soviet law, and while no one yet spoke of independence, the powerful leader of the USSR was definitely unable to impose his authority in the Baltics. Even the army was a doubtful resource in a country where everyone—recruits, reservists, their families, veterans—wondered and intended to decide for himself where his duty lay. Although the USSR had for decades been able to impose its will on people when rebellions occurred in Budapest, Prague, or Warsaw, no one in the USSR could express any doubts aloud. On Soviet territory, individual doubts and rebellion could not be transformed into a common awareness and thus play a political role. But the right to the truth, the public expression of that truth, and the USSR's acceptance of Eastern Europe's rejection of its system and tutelage put the finishing touches to the change begun in the national organizations. Public opinion blossomed under glasnost, and the people, framed by the popular fronts, took note of the reality: the system was splitting itself, however imperceptibly, while new ways of thinking and new types of authority slipped into place. The Baltic states gained their independence before it was proclaimed.

The second rupture, which happened in Russia, is in the long run more painful and more complex. Torn between its two callings—a European destiny and "the destiny of a hermit-nation," as Paul Claudel said, "that is an autarkical nation, tempted to break off the cultural dialogue with Europe, tempted to remain suspended in time and to abolish the category of the future"[45]—Russia had all the more trouble choosing because political renewal entered this debate and was mixed up with it and because all the movements that spoke of the Russian future did so in reference to these two extremes. For a humiliated, desperate people who are told (by Gorbachev) that everything had failed in this century and by the non-Russian nations that it was responsible for this failure, what a temptation it would be to turn back on itself and its past. The economic development of Western Europe then seems rather unappealing, for it exposes Russia's accumulated backwardness and increases the humiliation and tendency toward self-flagellation.

But this temptation can diminish before the other trend marking the Russian rebirth. The desire to understand the catastrophe that struck the country and reduced it to moral and material poverty propels an extraordinary intellectual movement. If the beginning of the century was an age of silver in Russia, the close of the century might also merit this name, probably not because of the artistic quality of the works produced but because of the wrenching effort made to heal the spiritual paralysis crippling the country. Thinking men and women—the initiators of the Silver Age whom Lenin's revolution drove out of Russia and out of minds and memory—are bursting forth in their country. As the overwhelmed country discovers philosophical ideas, a literature, paintings that had been hidden from it, in this burgeoning of books and ideas it may rediscover its identity. Today, the liberal intelligentsia in Russia is passionately devoting itself not to producing its own works but to encouraging the country's moral reconstitution. It can play a decisive role in achieving this goal. It opposes isolationism and destructive nostalgia. It wants to lead Russia to Europe, but because it is viscerally attached to its people and a humanistic idea of its destiny, it is trying to build a bridge between its European destiny and its leanings toward isolationism. The brilliant group gathered around Dimitri Likhachev—the Averintsevs, Tsipkos, Nuikins, Seliunins, Latyninas, and many others—knows that Russia must be both Russian and European. It must know how to be itself without yielding to the demons lying in wait for it in both camps.

For the Russia that does not yet know what choice to make and wavers between modernity and distinctiveness, the words *sovereignty* and *independence* seem as attractive as they do to the other nation-republics of the USSR. In Russia, however, they cannot have the same meaning or scope. If the people of the Russian empire can decide—even if that means thinking about the wisdom of such a choice—to leave Russia and the USSR, Russia cannot leave itself, or turn its back on the people of its empire, for the USSR is part of its destiny. It must free itself of the system that is responsible for its failure and for its turning into a wasteland and resolve the problem of its empire in an equitable fashion. This means that it must give people who wish to leave the freedom to do so, but it cannot get rid of those who wish to remain. The empire has become a burden to them, and the most backward part of the empire, the one that cannot leave it—the Muslim periphery—is dragging it into underdevelopment. Faced with the example of the other nations, Russia can not decide to move by itself toward progress, for it includes within its borders, in an inalienable Russian space, a large number of small nations and ethnic groups with which it is fated to live forever.

The Russian nation-state, a necessary stopover on the road to modernization, cannot resemble any other state. The huge problem of defining the future, separating from the empire while accepting itself as a heterogeneous, multiethnic country that is open to other cultures, explains the violence of the opposition within renascent Russian nationalism. It also explains the desperate effort of free minds—from Sakharov to Solzhenitsyn—to find a middle way that will prevent a frustrated society from responding to the appeals of extremists. Everyone knows that Russia must, like the other nations, free itself of the Soviet system, but it cannot deny its history without denying itself. On the other hand, the people it has dominated can free themselves from a country they have been forced to live with, but Russia can only reconcile itself with itself.

The end of the federalist system founded by Lenin calls for Russia not to recreate the logic of communism in some other form. Modernization is a reconciliation of societal and international life. If the class struggle and class hatred embodied by the Soviet system are replaced by the struggle between nations, then Russia will once again be suspended in a position that rejects modernity.

At the close of the century, the stakes of independence are in fact rather simple for the people of the empire but frightfully complicated for Russia. The vast Russian space now confronts a vast number of choices.

Part Four

After the Empire

10

From the Federation to the "Common House"

THE interethnic violence that ravaged the USSR beginning in 1988 sounded the alarm for leaders who were accustomed to thinking of the national question as secondary. Worried about the rising cacophony of unrest in the Caucasus in February 1988, Gorbachev decided to open the debate within the Party. It would, however, take him some time to realize the gravity of the national problem and find solutions other than an incantatory speech on the value of internationalism.

The debate within the Party took place finally in the fall of 1989, a year and a half after it was scheduled. During these eighteen months the intellectuals supporting perestroika tried to alert Gorbachev to what they were now seeing as the primary obstacle to any general progress:[1] "The Soviet government's policy is ethnocidal." "The Russian people are acting like bosses toward the nationalities."[2] These were futile warnings or nearly so. Gorbachev was mainly preoccupied with consolidating his power and gave only half his attention to the growing crisis. At the Nineteenth Party Congress held in June 1988, where he proposed remodeling the political system, the national question was mentioned, but in general terms: it was necessary to be equitable toward all the nations. The plan for political decentralization that he defended at the conference could open up areas of responsibility in the nations, but within what framework: that of the national states or that of the regions, by weakening the national state? No one was specific. The revised constitution of December 1988 showed that Gorbachev had not really included the national question in his concerns.

The Constitutional Provocation

On October 22, 1988, the Soviet press published the plan for an amended constitution and submitted it for public debate.[3] The republics immediately rebelled. First, the Balts and the Georgians—although they were not alone—vigorously denounced a plan of centralization that ignored all their demands and constituted a real step backward in federalism.[4] At a time when the nationalities were starting to organize and use violence or peaceful debate—depending on whether they were in the Caucasus or the Baltics—to show that the entire federal system was obsolete, this constitutional plan represented a real provocation.

A revision of the constitution was called for as soon as Gorbachev proposed to establish a genuine parliamentarism and agreed to define how it would be expressed by the Congress of People's Deputies and a transformed Supreme Soviet. Alas, it was precisely in the definition of the powers of the institutions designed to promote democracy—Congress, Supreme Soviet, Constitutional Control Committee—that the republic found that democratization would take place at the expense of federalism. On several points, their outrage was particularly well founded.

First, even though the right of secession had always been purely theoretical, this right, which was inscribed in the Soviet constitution, was basic for all the republics. It meant that under certain circumstances they could determine their own future. This right, in a truncated federation, was the symbol of a possible evolution toward an authentic federalism. The constitutional plan of 1988 practically annulled this right by giving the Congress of People's Deputies the exclusive right to "decide the composition of the USSR and ratify the forming of new republics" (article 108, paragraph 2 of the plan). Even though the right of secession was retained in article 72, it had little meaning compared to the total authority of the congress in this matter, especially since the makeup of this congress was unfavorable to the republics.

In the earlier constitutions of 1936 and 1977, the Supreme Soviet, a pseudoparliament, was a bicameral legislature in which the nations were represented twice: in the Soviet of the Union, as territorial voting districts—one deputy for every 300,000 inhabitants—and in the Soviet of the Nationalities. Each house had 750 deputies. The 1988 system overturned this balance by providing for a tripartite system of representation: 750 deputies for the voting districts, 750 for the national voting areas, and 750 for social organizations (parties, unions, and so on). This division, even though it ended up more or less equitably representing the nations within the parliament, was symbolically disastrous. The representation of the nations, strictly speaking—that of a multiethnic country of the federal type—fell from one half to one third. To the nationalities this meant that together they had the same status as the social organizations and that federalism was no longer on the same footing as the Soviet community.

Thus, the statutory decline of representation in the congress was accompanied by a decline in the representation of the republics in the second house of the new Supreme Soviet, the Soviet of the Nationalities, in favor of the republic of Russia. The constitutional plan reduced the size of the Supreme Soviet from 1,500 deputies to 542—that is, 271 per house—and thus modified the share granted to each national group. Article 111 granted seven deputies to each republic (instead of the previous thirty-two), four to the autonomous republics, two to the autonomous regions, and one to the national districts. The result of this division was that Russia, which comprised sixteen out of the twenty autonomous republics, five out of the eight regions, and all of the national districts, was much better represented relative to the republics than in the past, since, instead of benefiting from 32 percent of the seats in the Soviet of the Nationalities, the new system granted it 43 percent. Certainly, Russia represented 50 percent of the population of the USSR, but who, at a time of growing national demands, would accept a cutback in the representation of the nations? This was a very undemocratic way of taking them into account.

Two other reasons for the nations' discontent were the definition of the Supreme Soviet's power and the makeup of the Constitutional Control Committee.

The Supreme Soviet, or rather its presidium, of which Gorbachev was the president, had the exclusive ability (article 119, paragraph 1) to decree a state of emergency, which the Balts judged, in the event of a serious crisis at the periphery, allowed for the replacement of the local power by the central authority. Similarly, the definition of the Supreme Soviet's jurisdiction over economic matters—notably, prices and salaries—encroached on and even abolished the authority of the republics. Even though the economic autonomy of the republics was evoked as a legitimate way to satisfy their demands, these constitutional reforms of the central governing bodies confirmed that no one in Moscow contemplated playing the card of genuine emancipation.

Lastly, there arose a particularly sensitive point: the makeup of the newly created Constitutional Control Committee (article 125), which, beyond its president and vice president, was to have thirteen members. Who would they be? What role would the republics have in it? The vagueness of the statute led the republics to think that this institution, like the others, could be sullied by the same desire for centralization.[5]

This plan for constitutional revision contained some surprises. It was at variance with the current circumstances of the USSR, in which the national crisis was overt; with the whole official discourse of democratization and decentralization; and with the commitment to a greater respect for national rights. Did someone in Moscow wish to set off a political crisis at the periphery? Here, the convenient explanation of the tensions between the reformers and the fierce supporters of the status quo was particularly irrelevant: in fact, Gorbachev continued to prove remarkably indifferent to the complaints on the periphery. At this time (November 1988) he still conceived of his political reform in terms of the central government; for him, the nationalist problem was not yet one of the political problems of the USSR. It was clear that, concerning the political balance between the center and the periphery, this reformer was an advocate of changing nothing and hence eminently conservative.

There, encountering nationalist wishes, he recorded his first defeat. The outrage of the republics, especially the ones where violence had not supplanted political discussions, which gave their position much weight, ended up defeating the constitutional plan. The plan was amended, and the statute adopted on December 1, 1988, took account of their many objections.[6]

First, the national representation of the republics went from seven to eleven seats, and article 125 assured *all* the republics of representation on the Constitutional Control Committee. In a state of law, the constitution and the control of the constitutionality of the laws had to play a decisive role in the evolution of the USSR. It was well understood that the republics attached inordinate importance to the fact that the Constitutional Control Committee was federal in makeup.[7]

The republics also succeeded in obtaining the safeguard that a state of emergency could not be declared without consulting them and that no central administration could supplant or replace the local authorities.

Lastly, the right of secession was retained in its earlier form without any particular jurisdiction of the Congress of People's Deputies to limit its exercise.

On the whole, the demands of the periphery were heard, and the Supreme Soviet of the USSR adopted the amendment to the constitution by a wide majority (1,344 votes for, five against, and twenty-seven abstentions).[8]

The near-unanimous agreement involved not only the modifications made to the initial text but to the central government's commitment later to complete what was still only a sketch of the future USSR's political system. The second stage would involve the definition of the republics' areas of jurisdiction, which were to be considerably enlarged. Such, in any case, was Gorbachev's commitment, which for its part the Central Committee had already made on the eve of the voting.

At the time of the debate, Gorbachev recognized that the government had underestimated the situation, but he had since heard the warning given him. For all that, he was careful not to admit that the constitutional revision had initially been an attempt—a failed one—to increase the central power and, conversely, to reduce the authority of the republics. Did this new attitude indicate his new awareness of the agitation at the periphery and its causes? Or did it still simply reflect his traditional conception of federalism and represent a temporary concession to a national feeling that was condemned by history to come to an end? The political elites at the periphery had no doubts about the answer: revising the constitution was a deliberate attempt to stem the progress of nationalism by a process of accelerated centralization. Nevertheless, considering how readily Gorbachev acquiesced, it is tempting to choose the second explanation and surmise that at this stage the government still had not realized that federalism was dying in the national consciousness and that the status quo could not be preserved.

Despite provocation and negligence, the main issue still lay elsewhere—in the fact that on December 1, 1988, the nationalism at the periphery made the central government back down. It could be observed that self-contained confrontation, without any resort to violence, had completely changed the relations between the center and the periphery.

This first political crisis between Gorbachev and the republics had serious consequences. The personal authority of Gorbachev, architect of the revision, was already affected by it, as was the people's confidence in him, since he was blamed for the sponsorship of the initial plan and hence a centralist concept that belied his soothing words. Everything in fact suggests that Gorbachev, although he realized that he had to yield, still had not changed his conception of the national problem or understood what the outcome of the conflict implied for his authority.

Strong Center, Strong Republics

Curiously, 1989, the year of great political changes in the USSR—elections and the discovery of the value of parliamentarism—was the year of a real national stagnation in Gorbachev's policy. This stagnation is all the more surprising as it was accompanied by a radical shakeup of the national landscape—widespread violence in the southern USSR and a peaceful advance toward independence in the west. In this revolution of the nation in which everything was giving way—public order, central authority, chances for coexistence among ethnic groups—the intellectual and political immobility of the central government was astounding. In Moscow, it seemed that the only response to the breaking wave of nationalism was repression or disregard.

After an eighteen-month delay, the Communist party finally met to debate the problem on September 19, 1989.

Just before this meeting, the definitive version of the Party's "platform document" appeared in the press and presented the official position.[9] This document was new evidence of the intellectual paralysis afflicting the leaders of the USSR before the national question. The Party's statement, long overdue considering the tumultuous course of events, indicated no change in position, except toward Russia, whose awakening made the party take notice. This statement did not respond to any of the problems posed and indicated no progress toward true federalism, even though the USSR had evolved well beyond this demand. It contained every ingredient for further antagonizing the nation-republics, and assumed that the republics were still paying attention to the central government's positions.

The statement of the Central Committee, which was reminiscent of the debates of 1922 between the supporters of a confederal solution and the "autonomists" (that is, centralizers such as Stalin), concluded that Lenin's choice of federalism had been a wise one and that this conclusion was just as valid in the present as in the past. That is not how things stood in 1989, however. The only solution that might keep the USSR from coming apart would have been a transition to a very elastic, treaty-based confederation of sovereign states. This was what Andrei Sakharov persistently advocated.[10] And this idea was echoed by the most politically mature people of the periphery, including the Balts and the Ukrainians. The Communist party considered it unnecessary to make the slightest change in the constitution of 1922 and contemptuously rejected this confederal solution. Why renegotiate treaties since the constitution of 1922 was based on a treaty? Clearly, the Central Committee was forgetting, among other particulars, that in 1922 the Baltic states were independent and that in their case nothing attached them to the federation. The Baltics remembered this some months later when they took the step of declaring their independence.

Recognizing that past deviations had disturbed the smooth running of the federal system, the Party defined the system and the means to make it run. According to the document, the union was a voluntary association of states that, within one single state, the USSR, kept their total independence. In order for the federation not to be led astray as it had been in the past, it needed to combine "a strong center and strong republics."

It seemed unlikely that the republics would be satisfied with the reaffirmation of a system that they repudiated and with this altogether contradictory statement associating a powerful center and powerful republics. The republics also criticized the term used to define their sovereignty since it was the weaker of the two words that could be used: here, *samostoiatel'nost',* meaning "be master of its own destiny," was politically less precise than the term that the republics gradually took up, *nezavisimost',* literally meaning "not dependent" and explicitly excluding any other authority.[11]

Nor could the republics consider progressive the Party's insistence that Russia play a decisive role in the general progress of the federation. It was clear that the Party was more responsive to Russian grievances and the rise of Russian national feeling than to developments in the other nations. To the Russians, the Party sent a real message in this statement: it was time that Russia, like the other republics, had all the political and cultural institutions that allow a nation to thrive.

Lastly, a final point could legitimately disturb the large republics and the nations that had their own state. The Communist Party championed the rights of *all* the national minorities and suggested that the autonomous republics and regions become more important. For the nations with states, the suspicion dawned that the Party was employing its old strategy of weakening large groups by pitting small ethnic minorities against them.

These proposals were a provocation for the Georgians in the face of the demands of the Abkhazians and for the Azeris, who did not wish to acknowledge the Armenians of Karabakh. In fact, the republics interpreted this discourse as a return to, or perpetuation of, the well-known strategy of catching them in a stranglehold between Russia and their minorities while reaffirming their rights.[12]

Gorbachev's speech at the national meeting did little to dispel these suspicions. On the contrary, he insisted that the federation contributed to each nation and that indissoluble ties had been formed between them by economic interest. According to Gorbachev, self-determination was the principle behind the system, but the term had to be taken in its true sense—not as an encouragement of irrational secessions, but as the principle of self-management enabling each nation in the great Soviet whole to complete its full development.

At this plenum gathering, Gorbachev also gave his full support to advancing the Russian language as the state language of the entire USSR *(obshchegosudarstvennoi)*. This perhaps was an appropriate plan for improving many things in the USSR, but in 1989, when all the peoples were denouncing Russification, it also sounded like a provocation.

Similarly, Gorbachev jarred many national sensitivities by mentioning that the conditions of the Baltic states' attachment to the USSR were "beyond discussion." On this point, moreover, he was inconsistent with himself. On the one hand, he took over the idea of voluntary allegiance to the USSR, but, on the other, in his speech he admitted the need to resurrect the truth about the events of 1940.[13]

This complex and occasionally incoherent speech was intended at one and the same time to damp down conflict, to preserve the existing situation by indicating that any other course of action would be dangerous and essentially impracticable, and to send a message to the Balts that they were going too far and to the Russians that they were being listened to. On the whole, this speech left the impression that Gorbachev still did not understand that the very existence of the USSR was at stake. He appeared stuck in an unchanged position while accepting a few verbal or actual concessions. He said, for example, that special authorities were needed to discuss the national problems and hence exorcise the ghosts of separatism. As the Party leader, Gorbachev also set himself up as the uncompromising heir of Lenin. As such, he rejected unceremoniously the desire of the local Communist parties to federalize themselves. Party unity was a principle that was not open to debate. Here again, Gorbachev discussed the problem as if in the nations, notably in the Baltic states, the national parties still accepted this principle. Gorbachev's vision was not grounded in reality. These parties were marching toward independence, and Gorbachev was conducting a rearguard action to which no one was paying any attention.

"A strong center, strong republics": the formula was to last a long time. Barely had the highest authorities of the USSR proposed this surprising program—whose chief supporter was Chebrikov—to the republics than a growing chasm was seen between the actual country and those who still thought they represented the legal country. At the periphery, the sudden recurrences of violence and the activity of the popular fronts certified that the republics meant to increase their strength by opposing, ignoring, and even weakening the center. The chasm was inevitably growing between a center solidified in its perception of the national problem and a periphery that continued to affirm its autonomist or separatist objectives.

The polls proliferating throughout the USSR showed that the people were wondering about the federation's future. One opinion poll conducted in Moscow—a heterogeneous city with residents from all the nationalities but with an overwhelming preponderance of Russians—showed that even there, 72 percent of the respondents favored the stability of federal institutions, 18 percent resented the need to grant broad autonomy to the republics, and 6 percent wished that the republics aspiring to secession could follow through on their intentions.

The idea of a reorganization of federalism was also gaining ground among the supporters of perestroika.[14] At this time, the leaders of the national policy were liberals and had serious doubts about maintaining the existing situation. The most interesting comment on this issue was made by G. Tarazevich, a Belorussian and president of the Permanent Commission of the Soviet of the Nationalities, who, discussing the Party's platform, affirmed that the centralization of the USSR was disastrous for interethnic relations. He added that his commission would propose laws for protecting the republics' prerogatives. A cleavage also appeared between a Party theoretician of the national question, the guardian of an orthodoxy denounced by everyone, and the practicians who were really confronted with it. Tarazevich lucidly described the concrete problems, which the Party had refused to mention, of the internal borders of the USSR and the jurisdiction of the multiethnic republics in this area.[15]

The Federation and the Imperial Presidency

Neither the Party nor the congress had made any real progress in devising a new policy for the nationalities, and events were racing along. The Balts were convinced that Moscow would not go beyond plans for economic autonomy, which, after all, was a subject generating profound disagreement within the leadership team,[16] and decided that they would have to go it alone on the path that Moscow refused to open up to them. In January 1990, there was a dramatic confrontation in Vilnius between Gorbachev and a crowd of Lithuanians.

This trip showed Gorbachev's physical and moral courage—which he needed to face a calm but fiercely determined crowd—and was also poignant, for it revealed the central government's difficulty in reaching correct conclusions about a situation it finally was beginning to assess properly.

In Vilnius, Gorbachev finally took the step that he had long resisted. He recognized that federalism had never existed in the USSR, and he pleaded for a stay of grace, at last committing himself to a radical overhaul of Soviet institutions. After his return to Moscow, he addressed the Central Committee of the CPSU on February 5, noted that the rapidly deteriorating relations between the center and the periphery called for a reevaluation of the whole union, and, for the first time, suggested a possible diversification of the system of federal ties by adapting them to quite varied situations.[17] In mentioning this possibility, he was no doubt alluding to the platform—adopted by the Party in September—which said the exact opposite; but it was not inconceivable that, by noting an important turning point, Gorbachev felt it necessary to reassure the Party and persuade it that he still backed the program adopted shortly before. In any case, Gorbachev—for the first time—agreed with the many leaders of the Baltic states who pleaded for a new union treaty that would acknowledge the historical situation and developments that had already occurred there. These leaders said that a treaty that set up a confederation (no longer a federation) would have the advantage of sparing the USSR secessions that were already predictable. At the same time the journal the *Soviet State and the Law* proposed a similar analysis of the possible solutions to the crisis.[18]

The way suddenly seemed open to an initiative from the center. By proposing a new treaty to the republics, Gorbachev forced them into a dialogue while they clung more stubbornly every day to their own vision of independence. Was the USSR going to change?

It changed, certainly—but not as expected. The renovated federation then faded into the background because Gorbachev was mainly preoccupied by the introduction of a presidential system.

The Congress of People's Deputies that met on March 12 and 13, 1990, responded to this revolutionary reform, which had been made public a few days earlier.[19] The establishment of a presidential power had a strong influence on the rapid evolution of the national question. Its full scope had not been sufficiently appreciated, which explained the Lithuanian decision to opt for immediate independence without prior negotiations. Gorbachev, who had long rejected the idea of such a transformation of the Soviet system, accepted it—at least this was the justification that was given—in order to eliminate once and for all the threats of his expulsion from within the Party and to provide himself with the means to impose his reforms.

The presidential powers, especially as initially defined, were considerable, and like the plan for constitutional revision of the winter of 1988, they immediately triggered conflict with the republics.

The president of the USSR gained all the powers of the Presidium of the Supreme Soviet and its president. Because this presidium included the presidents of the Supreme Soviets of the federated republics, the change of March 13 meant that the republics were stripped of their authority in favor of the central government. This was not all, however. In the initial plan, the president of the union had the right to declare a state of emergency or martial law over the whole territory of the USSR. In December 1988, however, the republics battled to share this right with the congress, and their victory seemed to cast doubt on these extraordinary presidential powers. Finally, the president also had the prerogative of naming the members of the Constitutional Control Committee.

The debate at the congress no doubt forced Gorbachev to yield on two but not all points that were decisive for the republics. His two important concessions concerned the state of emergency and the Constitutional Control Committee. The president's imposition of martial law or a state of emergency would require the consent of the authorities of the republic affected by this decision. If this consent was not given, it could be authorized by a two-thirds majority of the Supreme Soviet of the USSR. In the same way, Gorbachev had to give up the right to name the members of the Constitutional Control Committee. What the republics had received in 1988 was thus once again recognized.

It still remained true, however, that the presidential system—even when accompanied by more balanced provisions—was not headed toward the genuine federation that Gorbachev committed himself to during his trip to Vilnius, and still less toward a confederation. This strong presidency was a means of centralization and probably could not be reconciled with the "strong republics" and with the union treaty that provided for differentiated legal status for the republics.

Although the debate forced Gorbachev to relinquish his nearly total power, his supporters were satisfied because, for them, what was foremost was that he was not eliminated by the opponents of change and that his policy of reforms could be pursued. His critics, however, saw the contradiction between Gorbachev's talk about decentralization and national autonomy and this concentration of power in his own hands. Boris Yeltsin clearly underscored how this agenda—first establishing a presidential power and only next reforming federalism—was inopportune and harmful to the calming of the crisis at the periphery.[20] A similar position was supported by Yuri Afanasiev, who could not be suspected of opposing Gorbachev in order to further his personal ambitions. He declared that first the constitution had to be modified and only then could a presidency be created. Furthermore, he said, electing a president in a multiethnic society had to be done directly by universal suffrage; without an election, the president could not claim the confidence of the nationalities. This reasoning was all the more judicious since the electoral system of 1988 had, as might be expected, considerably increased the representation of Russia: although it had 43 percent of the deputies in the old Supreme Soviet, it had 49 percent of the deputies in the congress elected in 1989.[21]

For the Lithuanians, it was clear that the powers written into the initial plan made it seem wiser to leave the federation before a too-powerful president could oppose them.

Despite the reluctance expressed during debates, the presidential system was adopted by 1,817 votes for, 133 against, and 61 abstentions. But Gorbachev's election to this tailor-made post was much less impressive; he obtained only 1,329 votes to 495 against him. The Baltic deputies boycotted this balloting, which, they said, no longer concerned them.

What clearly emerged from this political turning point was the opposition between Gorbachev and the nationalities. Gorbachev realized this, and after being elected, he plainly stated that he was committed to preserving the Soviet's state's political and territorial integrity and did not intend to preside over a dismantled state.

Although, given Gorbachev's choices, Lithuania adopted a radical position, the other nations of the USSR agreed to adopt a wait-and-see approach. They had little confidence in the soothing words and promises that Gorbachev sometimes lavished on them, since his actions—as exhibited by the choice of the presidential system—supported a centralizing, unifying, and authoritarian version of the Soviet state. No one disputed that reform as well as authority were needed. But which reform was the most urgent—federal or economic? Were the powers Gorbachev granted himself imperative for producing advances? The answer of the nations was clear: in resisting federal reform, Gorbachev, whatever his powers, was dooming himself to impotence. First, the nations had to reconcile themselves to the USSR by abolishing an unequal status that Gorbachev himself had recognized as unjust. It could not be admitted with impunity that federalism did not exist and indefinitely delay its reform. What Gorbachev lacked was not power, declared Afanasiev during this debate, but confidence. The nations reached the same conclusion. And the establishment of the presidency proved a fatal blow to the confidence that Gorbachev so movingly asked for in Vilnius.

For him, the presidency may have represented a great personal victory, but in the tormented history of the relations between the center and the periphery, its establishment represented a missed opportunity and, even worse, a bad blow for union. It provided further evidence of Gorbachev's extraordinary persistence in underestimating the seriousness of the national breaking point.

The Law of "Nonsecession"

Lithuania's independence, proclaimed just before the adoption of what was seen at the periphery as an "imperial presidency," immediately presented the new president with the national problem. This time he could not ignore its urgency. What could be done when a republic decided on its own to be self-determining? The silence of the official texts—the constitution and the Soviet laws included no provisions covering the exercise of self-determination—was on Lithuania's side when it claimed to be acting consistently with the law. The alacrity with which the central government acted to fill this juridical void contrasts remarkably with the five years during which it had left the national problem in limbo.

Since the word *independence* was first used in the various republics in 1989, the press periodically mentioned the need to define its dimensions. In Vilnius, in January 1990, Gorbachev stressed that independence could not be attained without appropriate laws, which he then said were to be revised. In January 1990, however, the agenda of the Supreme Soviet provided only for a debate on this theme; nothing suggested that a law would be proposed and hastily adopted at this session, but with remarkable speed the law on secession was passed on April 3, 1990.[22] This hastiness may explain the adoption of such an awkwardly worded statute for governing relations between the center and the periphery.

In the republics, which often greeted serious events with humor, this law, entitled "On the Procedure Connected with the Secession of a Republic from the Union," was immediately baptized "the law on nonsecession." This summed up the effect of a law designed to lay out a procedure strewn with traps that transformed secession into a nearly impossible undertaking, one impracticable for many of the Soviet republics.

The twenty articles of this law slammed the door on many hopes. The starting point for secession was a referendum coming from a republic's Supreme Soviet or a popular initiative representing a minimum of a tenth of that republic's population. Then a referendum by secret ballot had to be held six to nine months after the initiative with a two-thirds vote in favor of the proposal. This began a transition period of five years during which the USSR would intervene. During the initiative phase, the Supreme Soviet of the USSR was reduced to the role of observer except in the event of a serious dispute. After the referendum, it regained all its prerogatives. It had the right to judge the legality of the referendum and possibly call for a new vote in three months; then to submit the results to all the republics of the USSR for discussion; and finally, to present the results and discussion to the congress, which would debate them.

Only at the end of this process, whose duration was left unspecified, could the transition period of five years actually begin. This period was devoted to settling all the problems that the secession created for the rest of the union—economic interests, the USSR's military installations and strategic interests, the rights of persons, and so forth. Article 15 explicitly addressed the situation of foreigners resident in a republic who at the time of secession had to decide whether they wanted to remain. If so, the republic was then required to accept them. If they wanted citizenship, the republic was also obliged to grant it. If they wished to leave, the independent state must compensate them for the benefits they would lose and settle them elsewhere. At the end of this process, the congress, if everything was settled, would announce the secession. The parliamentary debate on the projected law confirmed that secession was then automatically granted and not given a qualified approval that must be submitted for further discussion.

Initially, then, the procedure seemed to have only one fault—its slowness. A minimum of six months was needed to arrange for a referendum, a few weeks or months to disseminate the results throughout the USSR and receive comments, and five years to adjudicate all the problems: in fact, that five years amounted to at least six, a very long time for the impatient republics. But this period required for separating from the USSR was nothing compared to the traps that riddled the procedure.

Chief among these traps was the two-thirds majority needed for the referendum that initiated the secession procedure. None of the republics had a national population that constituted two thirds of their total population, and the minorities or Russian residents were clearly disinclined to favor secession. This was true of the Ukraine, Belorussia, Uzbekistan, Turkmenistan, Azerbaijan, Armenia, Georgia, and Lithuania; six other republics were also below this two-thirds mark and would have great trouble convincing their electorate to approve such a plan. (Recall that the French law on the self-determination of New Caledonia required only 50 percent of the votes, which seems more equitable.)

More seriously, the law offered the minorities who had received a political status—autonomous republics and regions, national districts—to vote separately at the time of the referendum; their votes had to be taken into account independently. In other words, they could not only vote against secession but indicate their own plan, such as a secession within the secession of the republic and possible annexation to the USSR. It was clear that if, say, Azerbaijan went through this procedure for secession, it would instantly lose Karabakh, just as Georgia would lose Abkhazia, and so on. This arrangement was a major deterrent for the republics considering secession. In addition, even when secession was not demanded, it could only deepen the conflicts between the majority, who knew how much its minorities could threaten the dreams of independence, and the minorities, who would be tempted to abuse the power granted them by the law. No one doubted that, say, Azerbaijan, would try to repress the Armenians of Karabakh and would not care about the means it used to do so.

Because these arrangements did not apply to the republics that were administratively homogeneous, the law went further yet in the protection of minorities. It provided that where minorities without any special political status lived together as a group—as the Russians generally did—the separate vote would also apply. No republic escaped this. The law further stipulated that the populations foreign to the national majority had, like the minorities granted a territorial status, the right to choose their future—which for the seceding republic implied probable territorial losses that in many cases would constitute enclaves annexed to a bordering republic. The secession procedure specified by the law would then replicate the situation of Karabakh. No republic could calmly risk such a possibility. Instead, republics that wished to secede would be tempted, before starting the process, to use every means available to expel the minorities that would threaten its integrity.

But these draconian arrangements did not put an end to the ordeals awaiting the candidates for secession. In the final year of the transition period, if one tenth of the population doubted the benefits of separation, a new referendum could be imposed. Then, if the referendum did not win two thirds of the votes in favor of secession, another referendum could not be held for ten years. This was a highly unjust provision, but a prudent one from Moscow's viewpoint: as the Russians returned to their own republic and demographic trends rapidly changed the makeup of some republics, the republics that in 1990 could not amass the required majority could probably soon reach it. All precautions were thus taken so that a failed referendum would remain failed for some time and so that a winning referendum could be challenged.

An analysis of this repressive law reveals that the central government's sole purpose was to gain time in Lithuania. Moreover, from the day when Lithuania came out in favor of separation, Belorussia cited its own rights to part of its neighbor's territory: this demand could indeed be supported by the law of April 3.

Despite, or perhaps because of, its antisecessionist character, the law was approved in the two legislatures by a wide majority, with only thirteen votes against it. The Baltic states obviously did not take part in the voting, which they said did not concern them, for they could not secede from a union they had never joined. As for the deputies from the other republics, their votes can be explained by a certain indifference: this law appeared so inapplicable that, clearly, no one took it seriously.

Opposed to the law as soon as it was passed, Lithuania obviously objected to the law and did not consider itself bound by it, for it had been passed too late.

It remains certainly true that this law was the product of an error of judgment. All the republics saw this law of "nonsecession" as a trick and not arising from a genuine response to the national problem. Added to the establishment of the presidential system, it reinforced for the republics the idea that Moscow was not inclined to negotiate the dismantling or transformation of the USSR. What little confidence there was evaporated, and there was little left to squander. This law escalated the periphery's hostility toward Moscow.

The law also carried within it the seeds of new conflicts by aggravating the interethnic discord within the republics. Undoubtedly it helped Moscow to avoid danger. But the experience gained since the Alma-Ata riots showed that the disturbances at the periphery rebounded to the center and did nothing to help strengthen it.

Finally, the law was a mistake because it was inapplicable. Lithuania immediately proved this. When Gorbachev procured a temporary suspension of the declaration of independence, he promised that the negotiations between the Baltic republics and Moscow would be quick.[23] No one mentioned a delay of five years. The hypothesis can be suggested that this law was merely a smokescreen that allowed Gorbachev to arrange for Lithuania's departure without stirring up opposition. No doubt strategies of this type are common practice. They are effective, however, only when a political system has real authority and no one sees its successive relinquishments as signs of weakness. But Gorbachev no longer had any authority on the periphery. His enormous presidential powers did not enable him to impose his will in any way on the striking miners, the farmers who refused to rent land, the housewives who raided the markets, and so on. With Algeria, General de Gaulle had followed an ambiguous policy consisting of successive decisions that ended in independence. But the difference between the two situations is immediately apparent. The general's authority was real, and he had only a single problem to deal with; his decisions in Algeria were unlikely to affect the fate of other territories, for French decolonization was nearly over. On the other hand, the choices made by the Soviet government in the Baltic states were fraught with consequences for the whole or at least a large part of the empire. To the five years the government lost through an unwillingness to take account of the national problem were added decisions and laws that no one took to be definitive. At the periphery, a sense prevailed that improvisation had taken the place of politics among leaders in Moscow. If it was only a matter of separating from the empire, incomprehension on the one hand and skepticism on the other would be of little importance. Over the decades, however, shared space, shared human problems, and the shared desire for independence forced the central government to seek new formulas for coexistence.

Agencies in Aid of the Government

The USSR was indisputably making political progress, despite its delay in seriously dealing with the problems of the empire. One aspect of this progress was that no longer were any subjects taboo, as had long been the case with the national problem, whose reality had been denied. Discussion of the most complex situations overflowed from the narrow sphere of government to fuel a general debate. Within a few months, the national problem thus was turned over to various newly created administrative agencies for study. In politics the creation of commissions and committees of all kinds often is used to buy time, but in the USSR, where the people had long been cut off from political power, these commissions made it possible for people who often were outside the *nomenklatura* to participate in these discussions and hence to take up where the latter left off. Within the Congress of People's Deputies, a cluster of intellectuals constantly pressed for change, asked questions that distressed politicians who wished to maintain the status quo, and testified that here, as in Czechoslovakia, the contribution of the intelligentsia could prove decisive.

The earliest of these groups, the Permanent Commission of the Soviet of the Nationalities, specialized in national policy and interethnic relations. Presided over by the Belorussian Tarazevich, this commission was formed when the Congress of People's Deputies held its first session. Gorbachev said no problem was debated at such length and with such passion as that of the nationalities.[24]

It is notable that the representation of the nationalities among the permanent authorities of the new legislature and its new commissions was fairly equitable and balanced. Indeed, the presidency of the Soviet of the Nationalities was held by an Uzbek, Rafiq Nishanov, and that of the Permanent Commission, by a Belorussian. As soon as the Soviet of the Nationalities was more than a rubber-stamp assembly that met a few days a year and had become a permanent assembly working eight months out of twelve, the debate on the national problem clearly relied on representatives of a Central Asia that was hard to integrate and a Belorussia that formed an element of stability within the empire.

The Communist party, in a rare example of political insight, also understood the need to pay more attention to the explosion of nationalism. Its Central Committee created a Department of Relations among Nations, where representatives of the various nationalities worked side by side. Three sections—Non-Russian Republics, Russia, the Future—were to produce ideas that went beyond generalities. No doubt the Central Committee was seeking solutions of integration rather than the means of radical change, but the creation of this department already showed that, even here, the national question had ceased to be a "nonquestion."

Realism was taking giant strides. The new makeup of the Politburo, decided at the Twenty-eighth Congress in July 1990, gave greater weight in the Party to the national problem. The Politburo now was obliged to contain the representatives of *all* the republics, which was a total break with the rejection of representation of the republics characteristic of the Politburo during the early years of Gorbachev's tenure. But this quasi-federalization of the Party happened at a time when the Party, following the example of the federal state, disintegrated and was abandoned by the most prestigious luminaries of the Russian Communist party—Boris Yeltsin, Anatoly Sobchak, and Gavril Popov. In the republics, the popular fronts replaced the local Communist parties in public opinion as the authorities. It was without any doubt too late for the discredited Party's opening to the nationalities to have the slightest effect on their evolution.

On April 6, 1990, a new step was taken with the creation of a State Committee for National Questions, whose first function was to propose concrete solutions to the existing conflicts. Among its first tasks was to restore the rights of the national groups deported by Stalin. For the first time since 1945, the Tatars, the Germans, and the Meskhes had a permanent hearing for their grievances. The Council of the Federation was a new authority connected with the creation of the presidency;[25] the leaders of all the republics (each president of a Supreme Soviet) were required to join it.

Finally, the admission to the Presidential Council[26] of two national writers as representatives of their people who were as different as Chingiz Aitmatov and Valentin Rasputin—one a liberal and supporter of reforms and the other passionately Russian and conservative—testified to President Gorbachev's new concern to hear all varieties of opinion. In addition, Aitmatov was president of the Permanent Commission of the Council of Nationalities for Culture, Language, National Traditions, and Protection of the Historical Heritage. As for Rasputin, in the Presidential Council he wished to be the advocate for a cause that distressed the whole periphery—ecology.[27] Finally, the Ukrainian Grigori Revenko, who until 1990 was the first secretary for the region of Kiev, was given specific responsibility within the council for national questions.

This network of agencies and work groups united people from the republics that felt that the future of their relations with the USSR was their primary concern. These people had directly experienced the problems at the periphery and worked by collecting and comparing information about the most similar situations, as well as their own aspirations. In certain respects, the sudden flowering of groups specializing in national problems called to mind the beginnings of the Soviet government and Stalin's Commissariat for the Nationalities in which the entire federal system was once worked out. This could have been a sign of a desire to consider the whole problem, as was the case in the early 1920s—to work out, at a distance from the great debates of the Party and the parliament, proposals for a new policy. Or it also could have been a matter of creating smokescreens to hide the absence of plans and to buy still more time.

Whichever might have been the case, everything suggested that in the context of national revolution characteristic of the USSR, these groups were unlikely to be content with just supporting roles. People like Aitmatov or Rasputin were already too committed to the national debate to relinquish their power. Suggestions poured forth from all sides, moreover, as newspapers ran special columns and the scientific institutes organized round tables on this theme. A consensus was becoming clear: something of the former community of peoples of the USSR would be saved only if its radical differences of culture and level of development were recognized. Thus, the myth of a uniform Soviet people would be replaced by a multiform and differentiated community that could find its place in a common house with extremely flexible structures.

Various "forms of association" could be open both to those who wished to immediately take on the adventure of a federated common house and also to those who would be convinced by the experience of independence—which the USSR would have to ratify—that the states could live together only in solidarity.[28] Association, confederation of equal states, and so forth: many formulas were proposed. All or nearly all included Russia in this right to form an independent state; the flexible federating element must be distinct from Russia, including geographically. Russia would have its capital; the confederation or new "commonwealth" would have its own capital, as well.

In this conception of the future, the USSR gradually faded from people's minds. The general conclusion was that at a time when the people were coming out of their longstanding lethargy and becoming the agents of change, it would be inconsistent to ignore that the ethnocultural community was the natural framework of any society—the one where its solidarity was expressed and maintained.

11

A New People: The Displaced Persons

JUST after World War II, a new people appeared in Europe. They were given a new name, which has survived—"displaced persons." Refugee camps were set up to receive them, and organizations were created to help them straighten out their lives. The phenomenon gradually spread across most of the world—from the Palestinians who were not wanted in neighboring Arab states, to the people of Indochina who fled in make-shift boats and were dumped in camps in Thailand.

This was not an unfamiliar situation for the USSR. Stalin had transformed his country by displacing people: the peasants herded onto desolate roads leading to inhuman construction sites, the innumerable and anonymous people of the gulag, and later, whole ethnic groups deported. But at that time, when the USSR was no more than a huge concentration camp, a whole people of displaced persons, this situation was the product of a deliberate strategy. Society was to be broken up through mobility—of geography, status, and occupations—accompanied by terror and the total unpredictability of individual fates, for the purpose of giving birth to the society of the future.

As the USSR began to break up, as is generally the case during the dissolution of empires, it found itself once again faced with the problem of displaced persons, but this time not as the result of a deliberate choice.

This was a complex problem affecting various categories of individuals and a dramatic one because the fate of the displaced persons had to be determined within Soviet territory while the whole country was foundering in disorganization and scarcity.

Ex-Colonists: How Many Are There?

The first groups affected by the forced movements were the Russians and the Ukrainians, dispersed by the government over the decades to the four corners of the empire to help integrate peoples too far from the center to willingly accept its political line. The peripheral regions tended to see—or did so for a long time—these colonizing Russians and Ukrainians, who were actually drifting apart, as members of a single, Russian, group. After World War II Stalin and then his successors reserved a particular task for the Ukrainians, that of the second elder brother who shared the mission of extending central control and the Soviet political culture to the periphery. Long suffered in silence until the 1970s, this massive presence of foreigners assigned to represent the center in the republics was one of the earliest sources of discord between Moscow and the periphery. The voices of the intellectuals and their organizations were raised against the immigration of Russians, which was denounced as an attack on the republics' sovereignty. The arguments used against this policy of forced intermingling of populations were aimed sometimes at the centralizing political control, sometimes at the destruction of the nations' cultural unity, and sometimes at the Russian takeover of leadership posts and skilled or well-paying jobs.[1] By the end of the 1970s, these accusations became clearer: the systematic Russianization and Russian stranglehold on all the so-called technical or modern jobs led to the political, cultural, and economic dependence of the peoples thus administered.

Twenty years earlier, Khrushchev proclaimed that economic and social mobility in the Soviet Union was an established and voluntary fact that would eventually obliterate all differences and national prejudices.[2] Instead, however, it was noted that, from one census to another, the dispersal of the Russians and Ukrainians throughout the USSR isolated them and fueled growing national antagonisms. Even if the policy of the central government in this matter had not changed over the past thirty years, the Russians and Ukrainians themselves drew conclusions about the mounting hostility toward them and they moved away from areas where life was difficult to places where they found it more advantageous to live. From 1959 to 1989, the moving of Russians from one republic to another was sufficiently noteworthy to warrant collecting and analyzing the data (those of the Ukrainians are similar).

Table 11.1[3] shows that the proportion of Russians steadily declined in Russia because many of them moved to other territories and thus depopulated the republic of Russians in favor of other nationalities who settled there. This departure of the Russians from their own land contributed more than a little to their discontent and, in time, fueled arguments from nationalist movements for the Russians to return to their own territory.[4]

The Russian population, on the other hand, continued to increase as a percentage, and often in actual numbers, in the two other Slavic republics, Belorussia and the Ukraine, in Moldavia, and in the three Baltic republics. The Russian population declined throughout the Caucasus, Christian or otherwise, and in the Muslim periphery.

Under these conditions, it is hardly surprising that the data from the 1989 census concerning migratory movements within the USSR, which the central bureau of statistics delayed giving out, had been made public

TABLE 11.1
Percentages of Russians in the Total Population of the Republic

Republics	1959	1970	1979	1989
Russia	83.3%	82.8%	82.6%	81.3%
Ukraine	16.9	19.4	21.1	21.9
Belorussia	8.2	10.4	11.9	13.1
Moldavia	10.2	11.6	12.8	12.9
Lithuania	8.5	8.6	8.9	9.3
Latvia	26.4	29.8	32.8	33.8
Estonia	20.1	24.7	27.9	30.3
Georgia	10.1	8.5	7.4	6.2
Azerbaijan	13.6	10.0	7.9	5.6
Armenia	3.2	2.7	2.3	1.6
Kazakhstan	42.7	42.4	40.8	37.6
Kirghizia	30.2	29.2	25.9	21.4
Uzbekistan	13.5	12.5	10.8	8.3
Tadzhikistan	13.3	11.9	10.4	7.6
Turkmenistan	17.3	14.5	12.6	9.5

three months earlier in Estonia.[5] During a time of open national conflict, the Estonians wished to strengthen their case with an overwhelmingly obvious fact for the central government—the spectacular upsurge of the Russians in their republic. In 1959, only a fifth of the population was Russian; thirty years later, they were close to a third. This was the strongest possible justification for the Estonians' wish to close their borders to further immigration and to control their own population policy.

Over the years, the Russians left the Caucasus and Central Asia in ever-increasing numbers, a reflection of the mounting anti-Russian feeling in these republics. Although they left the places where they were unwelcome, the Russians did not return to their homeland, however. When they could, they settled in the Baltic states, which had a much higher standard of living than central Russia and to which the Russians were irresistibly attracted by the availability of Western-style consumer goods (clothing, furniture, high-tech appliances). In the Baltic states, the Russians had a sense of having already crossed the frontiers of their own country but in a setting that seemed less different from their own than what they had experienced in the Muslim countries.

The rush to the western part of the USSR was quite clearly spontaneous. Before Chernobyl, the Russian presence masked for a time the demographic disaster in the Ukraine and Belorussia, but since then this disaster has only become more pronounced.[6] Although the Caucasians and the peoples of Central Asia agreed that too many Russians were living in their lands and could not depart quickly enough, they knew that all they had to do was apply additional pressure—permanent irritants, demonstrations of hostility—to cause the Russians to flee. Things were quite different in the Baltic countries and the Ukraine, especially in the former; if the Balts did not wish to be engulfed by the Russians—already nearly the case in Latvia—they had in one way or another to close their borders to the influx of Russians.

The laws were not enough to halt this upsurge of immigrants that was favored by the central government. After all, where there were too many Russians, it was difficult to pass a law that would reduce their numbers or their rights. The Baltic countries were running a veritable race against the clock to check this invasion before it became too massive and thereby affected the population of the republic and its capacity for sovereign action.

Another goal of the proclaimed independence was to protect the republics from Russian migrants. Once independence was acquired, however, the Russians clearly appeared too numerous, too homogeneous in some ways, and their numbers not only had to be stabilized but also reduced.

Repelled by the Muslim countries where their life had become intolerable, the Russians were equally resisted by the republics where they wished to remain. Everywhere, they had to confront the same grievances that would eventually lead to violent conflict or large-scale departures. The titular nations intended to regain complete control of their political and administrative life. Thus, the Russian *nomenklatura* could not remain there. These nations also demanded the skilled jobs held by Russians. Although Central Asia had too few people who could fill these jobs, and although many of the posts abandoned by the Russians remained vacant and unemployment was rising, skilled replacements were plentiful in the Baltic countries, Georgia, and Armenia, and the departure of the Russians met a real economic goal. Finally, everywhere the remaining Russians were asked to cease displaying their cultural identity and accept the restrictive linguistic framework of the republics, which wished to be homogeneous. Accustomed to living in isolation and protected by the central government (of which they felt they were distant representatives) and bringing their language, habits, and behavior with them into their exile, how could the Russians adapt to societies that tolerated them only if they gave up any manifestation of Russianness? Could they accept being a people colonized by those whom they had colonized?

Like others before them, the Russians were soon forced to draw conclusions about the end of the empire, whatever the forms of cooperation that might succeed it.

This Russian ex-colonist population, whose numbers increased every year, had a high upper limit: 1,340,000 in the Ukraine, 1,341,055 in Belorussia, 1,652,179 in Uzbekistan, 6,226,000 in Kazakhstan, 339,000 in Georgia, 392,000 in Azerbaijan, 51,500 in Armenia, 474,815 in Estonia, 343,597 in Lithuania, 905,515 in Latvia, 560,000 in Moldavia, 386,630 in Tadzhikistan, 916,543 in Kirghizia, and 334,477 in Turkmenistan[7]—that is, more than 25 million people in addition to those who had already left without finding asylum in a republic other than their own. Close to 7 million Ukrainians also lived outside their own borders, with dense communities in Kazakhstan, Uzbekistan, and Kirghizia, not to mention those settled in the two other Slavic states and Moldavia, which allows us to estimate the number of potential displaced persons at about 30 million.

If the southern republics were closed to the Slavs, they would abruptly have to go into exile.

The Peoples in Search of a Territory

The peoples once deported by Stalin no longer had a home in the regions they were assigned to; at the same time, however, their desire to return to their country of origin encountered many obstacles. The Tatars of Crimea, who wished at all costs to regain the land of their ancestors, ran into opposition from the Ukrainians, who were assigned to the region after *their* deportation. Every trace of Tatar life has been obliterated there. The place names have vanished; public buildings and many houses have been destroyed. Despite this, the Tatars constantly returned to the attack, and although Moscow had long refused to listen to them, nothing ensured that in an explosion of intolerance, the Uzbeks and Kazakhs, who received 467,000 and 327,871 Tatars, respectively, would not someday force them to flee places where they would still be in exile. For the Ukraine, at the time of independence, did not intend to give back one inch of territory.[8]

After the recent riots in Ferghana,[9] a wave of panic gripped the whole non-Uzbek population in this region, and the Soviet press, which began closely following the migratory movements connected with the violence, revealed that the emigrants were clogging the means of transportation, making rapid moves nearly impossible.[10] Certainly, moving all one's personal effects is always an adventure in the USSR, but in the climate of fear that set in, complaints about the dreadful conditions of this exodus increased. It also was difficult for certain peoples to decide where to go. The Russians could possibly settle in Russia, but no one in the Caucasus wanted to hear about exiled Caucasians.

Another complex case was that of the Germans. After Germany became one of the USSRs favored political and economic partners, the Soviet government sought a new home for the German community of Kazakhstan. These people now felt little attraction for the banks of the Volga from which they were once driven. The Germans had formed communities in Central Asia, and although they met with no hostility, Kazakhstan had no intention of handing over any administrative territory to a non-Kazakh group. Was it necessary to send into exile or displace more than 900,000 Germans? And what was to be done with those equally numerous Germans who settled in Russia?[11] For a time, the idea of giving them the region of Kaliningrad (Koenigsberg) was toyed with.

Since the borders had been half opened for them, Jews continued to leave the USSR. Their numbers have declined from one census to another. This diminution was particularly spectacular in 1989, for they numbered no more than 1,367,910, down from between 2.5 and 3 million thirty years earlier.[12] Their emigration clearly was not due to a lack of territory. The Autonomous Jewish Republic—that was its name in the statistical records—was probably the most absurd example of Soviet territorial names. While the population of this region is increasing relative to the general average in the USSR, there were no more than 8,887 Jews out of a population of 214,000 on January 1, 1989, and they will most likely completely disappear from this area that still bears their name.[13]

The surprising contrast between the situation of the dynamic Tatars, fiercely demanding to return to their land of origin and the right to cultural and political autonomy, and that of the Jews, who have a territory that they have definitively abandoned, underscores the need for a serious and prompt revision of the ethno-administrative map of the USSR.

Meanwhile, whether candidates for relocation or for emigration from the USSR, these peoples added to the general anxiety and the material hardship of the displaced persons or those who were preparing to leave.

Finally, there was a category of Soviets whose displacement has been prevented or slowed down because the government had never been able to fully accept the truth: it was the population living in the areas irradiated by the explosion at Chernobyl.[14] The extent of Ukrainian and Belorussian territory affected went far beyond anything suggested; admitting it would lead to the opening of a new list of candidates for relocation. No doubt, whether Ukrainians or Belorussians, these Soviet citizens could be resettled in their own republic, which would not lengthen the interminable list of refugees whom no one knows how to manage and who find the doors of all the republics closed. As the truth was revealed, however, and the government recognized that new areas could not support additional populations, these people wanted to go farther away; they suspected that their entire republic would be forever dangerous. Half-truths, leaked at intervals to stave off panic and to control population movements, in fact had the opposite effect and ran a high risk of adding to the population exodus in the republics affected by the Chernobyl explosion.

What to Do with Displaced Persons?

Since the civil war in the Caucasus drove out the terrorized Russians—along with Armenians and Azeris—the problem became known and riveted the public, which recognized that a vast migration was beginning. Articles abounded, describing the immediate problem of the refugees in concrete terms.[15] Above all, however, the press wondered about the country's ability in the near future to confront the more general problem of settling millions of Soviets outside their areas of origin.

The short-term future was already worrisome for the Soviets, who discovered the government's inability to ensure some minimal reception for those whom the war had driven from their homes with no chance to prepare. Crammed into tents or sheds and receiving a tiny allowance (from 100 to 200 rubles) to meet their more pressing needs, the refugees from the Caucasus were, in fact, utterly abandoned and lived on what was provided by the surrounding population.[16] Letters to newspapers confirmed both the material distress of the refugees, the government's inability to deal with their influx, and the ambiguous attitude of those receiving them: spontaneous pity at first, but also fear for the future.[17] The hard life in the large Soviet cities could only become still more complicated if their population were suddenly to increase.

In the long term, the problem was almost insoluble and a source of serious conflicts. The refugees fleeing the violence—Russians settled at the periphery—had to leave in successive waves (in numbers still unknown) and often came from large cities where living conditions were relatively easy. When they were forced to leave, they wished to find a similar life in their homeland. Moreover, they thought that their homeland had incurred a debt to them. They had represented it in the frontier districts. They had ensured the system's permanence. Therefore, they expected the government would let them settle in the large Russian cities, preferably Moscow. Poorly received after their precipitate flight, they were asked to go where a dearth of population was holding up development:[18] to Siberia and the Urals, where the harsh climate and hard living conditions had deterred possible settlers. The higher salaries and often better housing there were inadequate to attract a permanent population. Once again, the government hoped to solve this problem by urging the refugees to settle there. The refugees, however, flatly refused because they felt they had been doubly the victims of their own fidelity—by ensuring a Russian presence at the periphery and then by being driven away. For them, the adventure should logically end in Moscow. But like most of the cities of central Russia, Moscow had neither housing nor jobs (unemployment was rising), and scarcity reigned.

In addition, the arrival of these refugees coincided with the return of the troops formerly stationed in Eastern Europe whom the USSR was forced to repatriate by the revolutions of 1989. It was impossible to take them in and provide them with living conditions that were like those they enjoyed in East Germany or Hungary. The Soviet leaders worried about what effects this difficult return of the troops would have on the army. The military commanders took a dim view of the hastily built camps. Gorbachev's whole policy was sharply criticized because it resulted in the hurried repatriation of an army that symbolized Soviet power to the outside world. Trying to forestall the army's discontent by arranging for the return of its men with the least possible inconvenience was a nearly impossible feat but one that was a top priority—which implied that there were even fewer places for those driven from the periphery by the rise in national hostilities.

The mass of Russians were unwanted in their own homeland but felt they had served it well in its distant possessions and surely were not ready to accept second-class citizenship and be forced into a new exile. Whatever the actual number of Russians rapidly returning to Russia, their demands were supported by those who still lived far away and knew they were themselves in turn threatened with having to leave on short notice. They therefore felt solidarity with these ex-colonists who were unhappy with a Russia that could not recompense them for services rendered. The Soviet government had to deal with and satisfy some 30 million Russians, but it was already overwhelmed trying to accommodate a half million soldiers. For the USSR, 30 million malcontents represented a major potential danger in addition to the threat to its equilibrium that the nations headed for emancipation represented.

But in Russia itself discontent was also mounting among the Russians who were worried about seeing already shocking living conditions further deteriorate under pressure from the sudden returnees. The never-resolved housing crisis reached catastrophic proportions that were now made public by glasnost and that heightened the general anxiety. At a time when the government ceased promising that everything would end well and that every Soviet citizen would have individual housing by the year 2000, the press announced that the repatriated soldiers, including some high-ranking officers, had to be content with "corners" in rooms that served as communal apartments. The train stations, in which people from the provinces were piled up in as though they were hotel rooms, now sheltered people fleeing with whatever personal effects they had been able to salvage from the disaster; Moscow increasingly had the look of a city in time of war. The limited empathy of the local population, which had already many times expressed its disapproval of people slipping into Moscow to settle there, was turning into open hostility. Renascent and nationalist Russia was also on the way to becoming a divided Russia, pitting the Russians of Russia against the Russians from the periphery.

What solutions could be suggested to resolve these growing tensions that were unforeseen only a short while earlier? The government could not forget that in the businesses around Moscow, pitched battles took place between Muscovites—that is, between regular residents with residence permits for the capital and the residents of the suburbs who wanted the right to settle in Moscow or who simply did it. The residents also clashed with the workers who had come from other republics, notably Central Asia, who had been urged to settle there in Brezhnev's time, when central Russia suffered from a shortage of labor. With increasing violence, the Russians criticized all of them for making unemployment and housing conditions worse, making the lines longer, and in short, being completely undesirable. In this atmosphere, any new influx of population risked provoking immediate conflict and, in any case, fueling profound hostilities. But what could be done with those who could not remain at the periphery? The government did not know.

Since 1917, the leaders of the USSR had been accustomed to maneuvering the populations as they liked, relocating them, and, regardless of human logic, carving up space according to purely political criteria, for the needs of the economy hardly entered into their calculations. "Humanity is our most precious capital," said Stalin. This capital was utilized most unceremoniously.

At the moment when, for the first time since the revolution, the government attempted to take the wishes of the people into account and become more democratized—the emergence of civil society testifies that it succeeded in doing this—all the problems connected with the brutal and often absurd actions of Gorbachev's predecessors came forth and compromised his own policy.

This policy encountered mainly human problems: the people resisted the implementation of certain indispensable reforms, and not applying them meant a growing catastrophe. In this difficult area of public opinion, the weight of some 30 million Soviets, for the most part Russian, who could not integrate in the country, risked proving particularly destabilizing and even downright explosive.

Although this situation represents an accumulation of problems brought about by earlier policies, the people who suffer blame those who are currently in power—accusing their initiatives of adding to the USSR's innumerable tragedies with the calamity of the legions of refugees of today and especially those of tomorrow.[19] The Soviet government and people have discovered both that the "Soviet people" never existed and that people whom no one thought about were coming to join others who already resisted the idea of living together—the displaced persons. This was a double unpleasant surprise.

Conclusion

Nation versus Democracy or Nation *and* Democracy?

AT the beginning of the 1990s, the question that was so dear to Lenin—"Who does what to whom?"—had its answer in the USSR and in the Europe that it had dominated for nearly half a century. The answer was unambiguous: national feeling toppled communism into the history of dead utopias.

The sudden defeat of communism under the combined blows of national communities determined to decide their own destiny first occurred in the USSR. That it started there was natural: communism began in Russia, and the Soviet Union that succeeded Russia set about imposing it from the outside; the general crisis of the system it embodied could come only from its weakening. The solidarity of peoples, that community of states ever willed and extended by Moscow—one day turned against the system and finished it off. Because the people of the USSR rose up and thereby eliminated all Gorbachev's chances of rebuilding his country's power, he withdrew behind his own frontiers and therefore had to tolerate everything outside it, being from then on intent to save what was most essential.

Originating in the USSR, objections to the imperial system thus reached the outside world and then boomeranged back to it, hastening the empire's breakup. According to an old proverb, nothing leads to failure as much as failures. The collapse of communism is an illustration of this.

We should keep in mind that the role of individuals in this collapse varied greatly. At the heart of the system, Mikhail Gorbachev, who presided over this extraordinary failure, was nevertheless an exceptional player. What he accomplished was extraordinary. He revealed that all his predecessors had plunged the country into a general disaster, and he had the courage not to rely on the condemned system to heal itself.

The choice of the man assigned with great responsibilities had an important influence on the course of events. In a few months, Gorbachev was able to give up his initial certitudes that the USSR could be quickly reconstructed, recognize the extent of the failure, and then attempt to democratize the country. He did not resort to force to preserve a condemned system for a time, although that was the kind of solution he had been trained for. He risked all his authority to force the USSR to take a good, hard look at the balance sheet of the previous seven decades and replaced the communist dream with an attempt at modernization.

Nevertheless, this courageous and generally clear-thinking man lacked something—an understanding of the phenomenon of the nation. In this sense, he was very much the heir of Lenin, for whom the nation, condemned by history, had only a strategic interest. By cleaving this way to an ideological position so contrary to his usual pragmatism, Gorbachev was reduced to passively witnessing nationalist convulsions that threatened everything he did and to observe the course of events without ever being able to redirect it.

Gorbachev was a great statesman who found himself overtaken by a flood of difficulties, and the absence of national leaders of equal stature is a historical oddity. During these decisive years, no nation of the USSR and no political movement at the periphery produced a true statesman. It is true that the ordeal of power reveals the stature of a Tadeusz Mazowiecki or a Vaclav Havel, whereas the struggle preceding their accession to power enabled them to show courage and imagination.

Nevertheless, no individual name stood out in the national struggles in the USSR. At the beginning of the century, when the nationalities of the Russian empire awoke, from all sides there came forth remarkable intellectuals, political men who were able to lead. Irakli Tseretelli in Georgia, Sultan Galiev among the Muslims, and many others played prominent roles in this history. Lithuanian independence, a still unimaginable experience in 1985—it seemed as unlikely as the clay pot breaking the iron pot—was managed by an unknown music teacher with no charisma whose unassuming personality was barely noticeable in the events he orchestrated, not unskillfully. It is doubtful that Vitautas Landsbergis will one day become mythical in the manner of a Lech Walesa. What he accomplished in his country is, however, no less remarkable.

With no heroes, no leaders of crowds, and no charismatic personalities the Soviet periphery rose up and changed the USSR and the European political landscape forever. Here the movements were born of the popular passion of the people. Their goal, the restoration of national identities, was more comprehensible to the people and closer to their aspirations than the imprecise—yet so urgent—goal of modernization that Gorbachev proposed to his country. Restoring the power of the USSR or modernizing it did not directly interest anyone, whereas restoring the nation was a goal that every member of a nation could understand and share. Gorbachev's tragedy lay in his perceiving too late what could urgently mobilize his compatriots. He will probably owe his place in history to the fact that, with a few exceptions, he avoided the use of force to stifle those national energies that he discovered too late and that neutralized his power.

Does the breakup of the Soviet empire mean that it will be replaced by a multitude of nation-states, each of which will set about shaping its own particular destiny? At the height of the family crisis of Soviet nations (and not of a mythical Soviet people),[1] no doubt all the nations, or nearly all, are thinking of a divorce, not imagining a compromise that would allow for the preservation of some links on new foundations. But the present intransigence of the nations cannot change the constraints of geography and history. The territorial continuity of the Russian, and then the Soviet, empire favored the coexistence of peoples of different origins and exchanges of their political and material cultures.

An empire that frees distant overseas possessions can take back its functionaries and colonists. Despite certain material difficulties, the USSR could also withdraw its troops and advisers from Eastern Europe and even possibly break all ties with its former allies. But where exactly would the borders of each component of the USSR or Russia be located? They have been moved so often in history, and whole populations dislocated, that to delimit the territory of each nation-state would be by no means easy. What is true of Russia is also true of the other states. What is the true border between Uzbekistan and Tadzhikistan? Where is the border between the Ukraine and a Tatar country demanded by the children of the people exiled by Stalin? Where is the one that separates Lithuania, which is oriented to Scandinavia, from Slavic Belorussia? These imprecise fields of rivalries between the renascent nations seem to be endless.

Economic constraints are no less important. Certainly, the ties of interdependence that Moscow imposed on the periphery—economic specialization, tangled lines of transportation, and the like—can with time loosen. After winning its independence, however, each nation has to see to its own survival, and this cannot be done by suddenly isolating itself from the whole on which it depends. Lithuania experienced this when it was subjected to a blockade, which, however, was far from being total.

There are only two solutions for disentangling the existing conflicts, avoiding those that are brewing, and settling the problems of interdependence—either the forced displacement of populations and borders, or compromise. The first solution is generally called for at the end of wars, in blood. Poland, at this end of the century, is a model of national cohesion, but until 1939 it was a battleground for national conflicts, and Jews, Ukrainians, and Belorussians living there once bitterly complained of being reduced to second-class citizens. The problem was resolved, if that can be said, only by the genocide carried out by Germany and Stalin's annexation of the western Ukraine. No one would opt for such solutions in the still-Soviet territory.

There remains a compromise that will enable different communities to live side by side in often multiethnic states. For this balance to develop, the nation-states clearly have everything to gain by fitting into a larger whole where particular conflicts would carry less weight. This explains why, beyond the discourse of rupture, the platforms of the national fronts and political parties include some thinking about the framework for independence. Should the three Slavic states regroup in an alliance among equals? Many Russians are thinking of this; the Ukrainian RUKH is not totally hostile to it, nor is the Popular Front of Belorussia. This solution would involve more than 220 million people and a huge amount of land and resources that would guarantee prosperity for the future. But where would the Russian land end? It clearly would not extend only to the Urals. Most of the Russian groups feel that it must reach the Pacific and probably include the whole northern part of Kazakhstan, populated by Russians and possessing a rich subsoil.

At the frontier of a Greater Russia or a community of Slavic states, could there be a juxtaposition of Muslim states of which Uzbekistan would be the leader? Or a unified Muslim state? Although political parties are forming on the Central Asian periphery—the ERK, for example, put forth such possibilities—in Russia, the idea of living next to a Muslim state with runaway population growth is frightening. It suggests to the extreme isolationists—those who continually deplore the "cost" of this periphery—that it would be wiser to find ways to compromise with it than to pit it against Russia. How can it be ignored that Central Asia has accused the USSR of having ruined it through specialization, deprived it of the means for future development, and condemned it to backwardness? The demographic dynamism of the Muslims and their political self-confidence compel Russia not to yield to the temptation to retreat to a uniquely European and Slavic perspective. Should the states of the Caucasus be abandoned to their hatreds, leaving Muslim-surrounded Christian Georgia and Armenia to an unpredictable future?

The former Soviet Union thus faces a serious dilemma. It clearly needs to bring the empire to an end, to inscribe its dislocation as a fact, and leave the peoples free to choose their destinies. In Moscow, no one can—those who wish to are probably few—refuse the nations that want the right to self-determination and separation. At the same time, however, if Russia, the heir of the USSR, wishes to avoid countless tragedies, it must imagine common solutions on a basis of equality for those whose destiny it shared. A new union treaty or federal pact is in the works.[2] Lively debates are taking place about what it should be; all forms are being considered.[3] This will take time, however. An interesting poll conducted during the Twenty-eighth Congress of the CPSU showed that of the delegates questioned about their vision of the future of the USSR, 70 percent favored keeping the federation, 11 percent favored a confederation, and 14 percent were for a combination of the two systems. This poll sheds light on the extent of the abyss between a conservative Communist party and a country in the throes of dizzying changes.[4]

What is definitely over is an empire organized around a common ideology (whether monarchy, Christianity, or communism) imposed on everyone by the illusory pretext of a higher power or historical necessity and continual progress. The USSR was the last empire of this kind, which gave it considerable might but also condemned it to turn its back on genuine modernization. Once the empire broke up, the nations—each of them—needed to decide the way to go in attempting this hitherto-failed modernization.

Clearly, the framework for progress will be the nation-state, alone or in various configurations: the aspirations of the peoples of this farthest part of Europe to strengthen the refound nation will not vanish overnight. The end of the czarist empire had once revealed that the age of nations had come, but the revolution put off its advent for nearly a century. This stage cannot be circumvented, however, without creating new disasters. Because it was humiliated for so long, national feeling cannot subside instantaneously. On the contrary, the new independence will give it life; reencountering old conflicts and unsettled disputes (those of territories demanded, dominations submitted to or resisted), it will create sharp oppositions between national interests. This has already been observed in Eastern Europe, where many old demons are returning, and it cannot be imagined that the nation-states born of the remains of the USSR will be spared.

Is this to say that nationalism is a regrettable development in an ex-communist Europe? That the only desirable development would have been the march toward democracy? That nationalism and democracy are mutually exclusive?

This idea, which reflects the contemporary experience of the great nations for whom independence and democracy were obtained long ago, overlooks the fact that the nation embodied in a state is a decisive stage in the process of a society's modernization. The nation in Europe, even in Eastern Europe, is not a tribe; the passionate wish to consolidate it in the framework of a state does not involve any slide into tribalism. Attachment to the nation—that is, the wider community knit together by territorial proximity and a common past and endowed with accepted social structures—is the distinctive feature of civilized humanity; it represents some progress over primitive society, not a regression toward it.

The nation builds itself around a common memory. It is natural, and even inevitable, that in linking up with this long-forbidden memory, the peoples who are freeing themselves also find in it traces of conflicts and animosities that have opposed them against other people. Recovering memory and paving the way to national cohesion is neither a simple step nor one free of suffering the separation of these things. In this hard process of reforging the nation, hidden tensions must return. This is a no doubt regrettable but inevitable aspect of a people's rebirth and discovery of their identity. Nationalism is too often aggressive and vindictive, but this is the product of a long period during which the nation was oppressed, humiliated, and even abolished.

Peaceful and open nationalism comes only later, when the certainty of the nation's ability to flourish has acquired the guarantee of time. Imagining that people could modernize themselves and so achieve a democracy by circumventing the nation and the powerful national feelings around which it consolidates itself is once again to reason as Lenin did and think that in the name of a simple postulate, certain stages in societal development can be bypassed. No one, after all, in the USSR thinks so, and, as the phase in which the national issue was ignored or underestimated is past, today the thoughts or plans of all those who have some political responsibility are centered around it.

Is this to say that the breakup of the USSR, which this explosion of nationalism produced, is unanimously held to be irreversible and that no reaction will set in to check or at least slow down the empire's fall?

The present situation of the USSR is surprising. In terms of military power, the Soviet state still has almost unlimited potential. Despite the disarmament agreements it has signed, its might is practically intact. A very large and superbly equipped army with considerable strategic potential could halt the country's dismemberment. In the short term, the end of the empire means a collapse of strategic power. Will those who embody this power—the state, the army, the KGB—passively witness its destruction while they still have the means for reacting?

In fact, aside from exceptional flareups—in Tbilisi and Azerbaijan—those who hold the Soviet power seem ready to accept their own liquidation. Is this happening voluntarily or because this very power has already decayed like the empire?

The second possibility is surely the most plausible. Although the military leaders have often vigorously expressed their disapproval in the face of the empire's breakup, they have not been unanimous. They have more and more sharply expressed the distrust they feel toward a multiethnic army that was becoming the battleground for national and even racial conflicts.[5] Such an unreliable army cannot be engaged in operations to restore order over the near-totality of the Soviet territory. Furthermore, the army as an institution has lost the people's confidence. Its leaders know this and say so. They have every reason to doubt that the people would accept the army's reestablishing order at the periphery through violence.[6] During the intervention in Azerbaijan, the people signaled their condemnation of this use of the armed forces inside the territory. The presidential decree of July 26, 1990, by which Gorbachev demanded the immediate dissolution of the national militias and threatened to use force against those who resisted, could rightly lead to skepticism: some efforts to use the mailed fist are certainly imaginable, but could they be carried out in both the Caucasus *and* Central Asia? How could so many hidden arsenals be detected? Who would care to undertake such a operation?

What a demoralized army, deeply infected by the virus of national oppositions, cannot and will not do, the KGB does not seem any more inclined to attempt. The result of this is that no force will probably stop the irresistible race of the nations toward their chosen destiny. The USSR's situation in 1990 called to mind that of Russia in the spring of 1917. At those two moments of its history, the state's power was still almost intact and yet was paralyzed before the forces that defied and were going to destroy it. In 1917, however, people and nations at war, ignorant of the ways and means of totalitarianism, allowed their revolution to be taken over by the Bolsheviks and could not resist the absolute domination imposed on them. At this century's end, the people of the USSR have a tragic knowledge of the totalitarian experience, and those who are leading the national fight—in Russia or on the periphery—have set about giving them back a memory and understanding not of Stalinism but of the profound sources of oppression. Is this not, for this developing national movement, the best way to approach democracy?

Epilogue

The Breakup of the USSR (1990–1991)

WITHIN one year—November 1990 to November 1991—the fragmenting USSR unequivocally came to an end. Finishing off Lenin's political system, the mighty Communist party, and the patiently assembled empire of the Bolsheviks required a coup d'etat—failed, of course—but also and mainly the unremitting hunger of the people belonging to the Soviet Union to determine their own fate. By November 1991, the signs of their victory were unmistakable. After more than seventy years, the traditional celebrations of the revolution were given up and replaced in Leningrad (now St. Petersburg) by ceremonies to mark the shedding of the name carried by the city since 1924, the year of Lenin's death. Elsewhere, ceremonies commemorating the triumph of the Bolsheviks gave way to demonstrations commemorating the victims of the Soviet regime. Still elsewhere, some nostalgic communists, with the red flag in front of them, maintained that the whole experience of 1917 had not failed. Nearly everywhere else, however, the red flag was replaced by the national flags that were reintroduced in 1989. And even over the Kremlin itself, the symbolic center of what remained of Soviet power, the red flag of the revolution flew in close proximity to the tricolored flag of Russia. In this competition of national emblems the red of communism increasingly capitulated to the brilliant, varied colors of the national traditions, a metaphor for the events in the huge area that is no longer Soviet land.

The Soviet Union evaporated, and the search for a new bond between the republics continued to feature the retreat of a central government that no one knew what to call.

What Union? Who Will Be Part of It?

In the USSR of the fall of 1990, the idea still prevailed that the nature of the Soviet state soon would have to be defined as would the relations among its constituent republics. On October 24, the Supreme Soviet of the USSR, created by the elections of 1989, passed a law confirming the supremacy of federal law over the laws of the republics and that this principle was inviolable, at least until a new union treaty was signed.

On that date, the situation in the USSR was complicated. All fifteen Soviet republics had proclaimed their sovereignty, beginning with Estonia on November 18, 1988, and ending on October 31, 1990, with Kirghizia (whose Supreme Soviet approved the declaration of sovereignty on the first reading). Russia, the federation's chief republic, so often confused with the federation as a whole, declared its sovereignty on June 11, 1990.

In addition, a few months later, deeming that sovereignty was only one element of national self-determination, four republics—Lithuania, Estonia, Latvia, and Armenia—advanced to the final stage, independence. The Soviet government certainly wished to discount these aspirations to exercise the right to secession written into the still-lawful Soviet constitution, but this rejection had no effect on the republics, which no longer considered themselves a part of the union.

The USSR was not the only victim of this riot of independence. Russia, which encompassed sixteen of the union's twenty autonomous republics, was suddenly faced with their rebellion and their claim to be doing to Russia what Russia had done to the federation. From August to October 1990, ten of the sixteen autonomous republics proclaimed their sovereignty, with two immediate consequences. Institutionally, they thereby intended to enjoy a status equal to that of the fully independent republics like Russia or Ukraine and depend only on the authority of the central government. Economically, they thought this implied that they would have complete control of, and the profits from, their natural resources. This claim was particular worrisome for Russia because it was made by the autonomous republics that produced gold and gemstones (Yakut and Buryat) and oil (Tatar).

Georgia had a long history of conflict with its minorities and was another republic bedeviled by these national movements. The Abkhazian Autonomous Republic rebelled against it on August 25, 1990. The Supreme Soviet of Georgia abruptly nullified this declaration of sovereignty but failed to induce the Abkhazians to give up their aspirations.

In some cases, some smaller territorial and political groupings—autonomous regions and districts—even decided that they, too, had a right to sovereignty. The most remarkable and destabilizing instance of this was once again in Georgia, where on September 20, 1990, the autonomous region of Ossetia proclaimed its right to become a separate republic. As in the case of Abkhazia, Georgia's response was blunt. Emphasizing its sovereignty, Georgia held that all its territory was essential to it. The Georgian authorities thought that the Abkhazians and Ossetians were merely pawns of Gorbachev and the other Soviet leaders who were using them to weaken Georgia and force it to comply with Moscow.

In any case, in the fall of 1990, the upheavals in the USSR were affecting life in the multiethnic republics. The growing anarchy clearly called for some minimum organization.

The debate over organization divided those in power. Three conceptions of the USSR's future then emerged and dominated all attempts to devise a union treaty. Still thinking he was in control, Gorbachev wanted to impose a union that was a federation. For him, this meant not only the preservation of a central power but a strong center working with strong republics. He accepted the sovereignty of the republics within the framework of the sovereignty of the federation, which was an integrative structure. Confronting Gorbachev was Boris Yeltsin, who argued for a confederation—even though he avoided the term—based on the sovereignty of the republics, direct agreements among the republics, and a center that was of course retained but deprived of any means for interfering in relations between the republics. Although Gorbachev conceived of the whole system from the center, Yeltsin organized it from the periphery, sidestepping as much as possible the center's involvement. The logic of Yeltsin's plans was strong republics, weak center. The third variant was the one proposed by the Baltic states: pure and simple separation, thus depleting the union of its components.

Added to this first debate on the future union was another one on the limits of the right to sovereignty. Could any national group assembled on some determinate territory claim sovereignty, or only the constituted republics that could generally claim to descend from older governing structures? Here, too, positions diverged. Gorbachev favored recognizing sovereignty in anyone who requested it. We can see the advantages of this idea. The minorities of nearly every republic in the USSR demanded sovereignty. By supporting the right of the Abkhazians to form a sovereign state or that of the Lithuanian Poles or the Estonian Russians, Gorbachev may have been hoping to squelch the desires for independence in Georgia and the Baltic countries by showing them that independence would inevitably lead to the disintegration of their territories. Confronting Gorbachev, Yeltsin and the leaders of all the independent republics were agreed in defending the idea of sovereignty limited to the constituted states. It was then up to them to find conditions of life together acceptable to their minorities. But Yeltsin said that the status of minorities within the republics was part of their sovereignty and could not be interfered with by the future union. After all, a sovereign republic must be master of its territory and population.

The profound differences between the various parties to this debate made it hardly surprising that the union treaty was the occasion for interminable negotiations punctuated with sudden electrifying developments and failures, and—in the last analysis—the first cause of the collapse of the whole Soviet system.

The first stage in this pursuit of an impossible union was the treaty plan prepared and presented by Gorbachev on November 23, 1990, and then submitted for public discussion. Worked out at the Council of the Federation in consultation with all the republics (except for the three Baltic countries and Georgia, which abstained), this text initially had a rather favorable reception, mainly because of its nonideological nature. The word *socialism* did not appear in it, while the condition of rights, the choice of forms of property, and the modern management of the state and the economy all written into it by the plan's drafters suggested that it expressed a new vision of the relations between the nations of the future union.

Nevertheless, when the plan was presented to the parliaments, it provoked considerable opposition. The Congress of People's Deputies, meeting for its fourth session in December 1990, witnessed a clash between Gorbachev and Yeltsin, a clash not dramatic in its form but serious in its substance. Although Gorbachev praised the plan—"We are through with the unitary state and have developed a multiethnic state"; "We have found equilibrium and coherence between the sovereignty of the USSR and of the republics"—Yeltsin counterargued that "the center is trying to hold onto its unlimited power over the republics"; "The center is ignoring the declarations of sovereignty and trying to preserve its bureaucratic control over them." The two men were no less opposed on secession and sovereignty than they were on the plan's basic orientation. For Gorbachev, existing Soviet law applied as long as the treaty and new law had not been worked out, and, of course, this referred to the whole legal system that made the center paramount. For Yeltsin, this was already an outmoded conception; the republics should decide what laws and principles applied to them, and the center was obliged to defer to these manifestations of sovereignty.

Meeting in the days that followed, the Russian parliament witnessed a renewed conflict between conservative deputies who favored the plan and democratic deputies, like Galina Starovoitova, who accused Gorbachev of presenting a thinly veiled variant of the 1922 Pact of Union. More generally, many deputies protested the use of the expression "union treaty" as too reminiscent of the old system. At this moment, Ukraine vigorously opposed the plan and refused to adopt any union treaty before working out and implementing its own constitution. Russia immediately took the same position. The advantages of this agenda were clear. Republics that adopted a new constitution independently of the central government would approximate genuine sovereign states, fortified with a basic law not subject to any higher law, namely the union treaty. It would thus be difficult to reinstate the old order—the supremacy of the Federal Constitution and the Pact of Union—over the national laws.

This debate shows that Gorbachev's position was to argue every inch of the way for the urgency of the union treaty, for its priority over any institutional construction, for establishing beforehand the mode of functioning of a sovereign state, and then and only then, for working out plans for the common life of its constituent states. The plan of December 1990 merely intensified an already heated debate over whether to build the union from the center or from the republics.

In the attempt to find a way out of this dilemma, a second version of the plan was presented in March 1991. The two versions differed little; opposition to it was nearly as strong and the treaty remained stalled.

One feature of the two versions made them unacceptable to all the republics—the one concerning the right to secede. For the first time in Soviet constitutional history, the amendments of November 1990 did not mention the right to secede. On the other hand, they did describe the conditions under which the member states of the union could be expelled from it. They also defined a bifurcated union—one made up of states that were signatories to the treaty; the other, of those who had refused to sign it but could not withdraw from the union and who remained subject to the Soviet laws in effect. These restrictive arrangements, in complete opposition to the country's real development—the Baltics had already proclaimed their independence—immediately condemned the sovereign federal state, the end product of these two successive plans, to being a mere idle fancy.

A new attempt to advance the debate was the referendum of March 17. Consulted about its desire to preserve a union, the people came out broadly in favor of union, thus giving Gorbachev the necessary support for working on a treaty. No doubt the reference to public support had some weak points. Several republics, beginning with the Baltics and Georgia, had banned the referendum from being held on their territory. The results also were tainted by irregularities in Georgia, for example, where the army had "protected" the voting in minority areas and the electoral lists and the verification of the voters' identity were questionable. More seriously yet, various republics had added their own questions or modified the ones submitted for referendum. Thus, although the vote was in favor of the union in the Ukraine, the western part of the republic came out for secession, pure and simple.

Despite these results, negotiations about the treaty were reopened. On April 18, 1991, a meeting was held in Kiev by the representatives of the five largest of the Soviet republics—Russia, the Ukraine, Belorussia, Uzbekistan, and Kazakhstan. Remarkably, this meeting of the Big Five—held in the absence of any representative of the central government (Yeltsin's idea of a "horizontal" government was manifesting itself here)—to forge a common position produced one outstanding result. The large republics agreed not to allow the autonomous republics to take part in the debate over the union. The Big Five thereby clearly indicated that they would not let the center weaken the large republics by playing on the little separatisms that then were threatened. Therefore, the general discussion became acceptable for all. Briskly run in a government dacha, it ended on April 23 with the signing of the Novo-Ogarievo Accord reuniting nine republics (Armenia and Moldavia had joined the three Baltic republics and Georgia in dissenting). This accord, concluded not on the basis of the (amended) plan of March 1991 but simply on the idea of the treaty's urgency, clearly posited that only the republics of the union would be signatories to the treaty. Decisive on this point, Yeltsin and his colleagues had beaten Gorbachev, who henceforth would at each stage of the debate have to make further concessions that even more greatly weakened the central government.

In a hurry to reach this goal before going to London in July to meet with the heads of state of the world's seven leading industrialized countries, Gorbachev tried—in vain—to impose his agenda and views on the negotiations.

A third plan, which Gorbachev presented over Soviet television on June 18, urgently had to be submitted, he argued, to the parliaments of the republics so that the treaty could be signed before the end of the month. This time, Gorbachev thought he had sacrificed enough to the ambitions of the republics to win their agreement. The republics had won their case on two points. The expression "sovereign states," whose meaning Gorbachev did not wish to specify, was retained: the autonomous republics kept their inferior status and were excluded from the treaty. On the other hand, the controversial question of the financial contribution of the republics to common expenses was not resolved in the way the republics desired. The latter intended to levy taxes on their territory and determine their contribution to the common budget. The plan submitted to the parliaments stipulated that, when the federation's means of subsistence was in doubt, it was up to the federation to authorize and collect taxes. Similarly the plan was opposed to the republics' wish to form their own national armies, and it held that defense came exclusively within the federation's province. Finally, the Constitution of the Union—for the plan confirmed its existence—had to be worked out by the authorities of the federation alone.

Once again, Gorbachev hoped to have most of his ideas carry the day in the plan, whose discussion and acceptance he tried to hasten by arguing for the need for his country to exist in a peaceful world and also by affirming that this plan harmoniously retained the center while conferring considerable powers to the republics. Once again, his hopes were dashed. First, he had deliberately ignored the points of contention several times pointed out by Boris Yeltsin. Furthermore, at the end of June he came up against a novel internal situation.

On June 12, despite the efforts of the entire Soviet leadership, including Gorbachev, to forestall this event, Boris Yeltsin was elected by universal suffrage to the presidency of Russia. This suddenly jeopardized the future institutions of the union, particularly the presidency. The president-elect of Russia was less inclined than ever to put up with central structures that would hamper his activities. The problem of the center was even more drastically presented by Ukraine, where Leonid Kravchuk, chairman of the presidium of the Ukrainian legislature, refused to sign a treaty that put the union and the republics on the same level. He said that the only union possible was not "nine plus one" but a union agreed to by the new signatory republics. The union, a flexible alliance, had no need of an additional partner who must in no way be a sovereign state. Russia was determined to downgrade the status of the union, and the Ukraine very much wanted a union, but without a state as such. In these circumstances, the agreement of the seven other republics, which after all, were not absolutely unanimous, carried little weight.

The Coup and the End of Federal Hopes

Until March 11, 1990 (the date of Lithuania's independence), the Soviet Union was made up of fifteen republics. At the time, the negotiations on the union treaty ended in a statement of the "nine plus one" type, and then in a group of seven republics more or less resigned to signing some sort of accord.

In August, these interminable debates and the permanent revision of the plan seemed to have borne some fruit. A significantly modified plan making many concessions to the republics was about to be adopted, although not by everyone. The Ukraine would not announce its decision before the end of the year, when elections would be held, and Russia intended to get many more amendments. On the eve of the signing, Yeltsin did not hesitate to declare at a press conference that he and Nursultan Nazarbayev, the president of Kazakhstan, wanted substantial changes in the plan that would even further reduce the center's authority. In fact, besides Russia and Kazakhstan, whose accord was conditional, only three other republics were ready to take part in the signing on August 20—Uzbekistan, Tadzhikistan, and Kirghizia. The others (minus Ukraine) promised to sign in September, but the number of potential partners in the union was constantly diminishing.

Gorbachev's main goal, however, was the adoption of the treaty and the survival of even a badly injured union. This hope was dashed by the coup of August 19, which destroyed what remained of the USSR.

Considerable numbers of books about the coup have appeared in the

USSR, and all suggest that one of the conspirators' main objectives was to act before the signing of the union treaty. In their minds, any attempt to check the erosion of the system and its structures had a preliminary condition—to maintain the general framework, that is, the USSR. On August 19 (the plan was to be signed on August 20), a compromise whose inadequacies the parties agreed to underscore was violently attacked by Lukyanov, president of the Supreme Soviet. The person later said to be the "ideologist of the coup" then strongly emphasized that the plan took no account of the critical remarks of the Supreme Soviet of the USSR, that it gave the republics untold opportunities to suspend the laws of the union on their territory, and that it was in the last analysis a bonfire fueling the "war of laws" and heating up the conflicts.

Meeting in a smaller group two days earlier, the USSR's ministerial cabinet expressed the same reservations. The central government's problems and constant postponements in signing a treaty and Russia's rise in power (which since Boris Yeltsin's election appeared to be the union's rival state, a rivalry symbolized by the installation of Yeltsin's offices in the Kremlin and the coexistence of two flags on the spot that since 1918 had been the center of Soviet power) were signs that henceforth the union's continued existence was questionable.

The coup's failure triggered a development that Gorbachev had long been trying to avoid. First, the coup prompted those who had already abandoned the USSR to consolidate their independence. On August 21, Latvia, which had adopted a plan for the restoration of de facto independence, voted for a declaration of total independence. Moldavia affirmed that it would proceed with its secession. In Estonia and Lithuania, the banning of the Communist party indicated the desire for the prompt elimination of all agencies that allowed the Soviet government to interfere in the life of the independent republics. Zviad Gamsakhurdia, the president of Georgia, appealed to the international community, emphasizing the urgency of recognizing the independence of the new state in order to avoid their overthrow by a successful coup. For the same reason, the government of Ukraine aggressively reverted to its favorite proposal—to set up national units and put Soviet troops stationed in the republic under Ukrainian control. In Armenia, the referendum of September 21, in which 94.3 percent of the electorate voted, resulted in 99.31 percent of the votes being in favor of immediate independence. Once the coup was over, it was important for all the republics to strengthen all means for resisting the central government and the union that would not quite expire. The idea was gaining ground that the union was dangerous and hence must cease to be.

Could a common plan be reactivated after the coup? On his return from his confinement in his vacation home in Foros, Gorbachev at first thought so. But reality—a legitimacy weakened by a coup whose perpetrators had been his closest collaborators, the suspicion of softness or complicity that dogged him, and the growth in prestige of those who mounted the resistance to the coup, with Yeltsin at their head—forced him to reconcile.

The new Soviet parliament organized after the coup met for the first time and grappled with the still-pending problem of the union. It was an odd parliament in which the governing structures were undefined. On September 2, 1991, the Congress of People's Deputies, meeting in special session, analyzed the failed coup and admitted that the whole organization of the hitherto existing state had vanished. Nursultan Nazarbayev, the president of Kazakhstan and now one of the country's leading players, then became the spokesperson for a radical political revolution. Noting that the constitution no longer existed, he proposed the setting up of transitional political structures in expectation of the much-discussed union treaty that would provide a definite spatial framework for the future system. Meanwhile, the congress, which did not want to commit total suicide, organized a remodeled assembly for concluding the union treaty as quickly as possible and, above all, setting up interim power structures in the State Council. This constitutional revolution led to a decisive role for the republics—made their presidents become equal to the president of the USSR in the State Council—and to the quasi-nullification of the central government.

Is the Economic Community a Substitute for Political Union?

In this context in which the republics' authority continued to be affirmed, an agreement was worked out for an economic community that Gorbachev still hoped would be the first stage of a union treaty. For many of the republics, however, it was the hoped-for end product of all the negotiations.

The agreement signed on October 18 was primarily the result of the efforts of Nazarbayev, president of Kazakhstan, and for several days, Alma-Ata, its capital, was the actual political center of the country. He endeavored to persuade his partners that because of a lack of agreement on a political future, first things came first and they had to find modes of a common economic organization. He finally managed to obtain agreement on a plan drafted by a team of specialists headed by Yavlinsky, the source of most of the plans proposed since the coup and whom the West considered an authority on economics.

Debated in Alma-Ata, the plan was finally signed in Moscow on October 18 by nine heads of state: eight for the republics and Gorbachev for the USSR. A close reading of this plan and comparison of it with the plan accepted on October 1 at Alma-Ata make clear the retreat of the center and the dynamism of the republics. Certainly, the signatories agreed, for want of a political union, to set up an economic community of "independent states, members or former members of the USSR." Nevertheless, with the exception of Armenia, the former members were careful not to take part in the preparation and not to sign the final document. In addition, Ukraine withdrew: "It has nothing to sign with the center," said the vice president of the national parliament, and at the last moment, Azerbaijan decided to stay outside the community. A few hours before the signing, Mikhail Gorbachev let it be understood that even the most independent states could join a community limited to economic matters, but he was soon disabused of this illusion.

Basically, this long document, consisting of fifty-nine articles, had been oriented in a direction unfavorable to the center on three fundamental points:

- The Gosbank (state bank of the USSR), which was initially to play a central role as the banking union of the republic being set up, was in the final version (article 18) immediately forced to share its prerogatives with the central banks of the member states in a Provisional Board of Directors;
- The budgetary community, for which Gorbachev had tirelessly fought, was certainly referred to, but in practice the federal budget was funded by payments (by the member states) corresponding to "sums fixed" by special agreement of these states (article 24); thus the federal budget depended entirely on the good will of the participants;
- A final cause of bitterness for those who were nostalgic for the federal

center: in the plan negotiated at Alma-Ata, all the articles that maintained common governmental structures had to be dropped from the final plan, which implied the disappearance of ministries common to the various signatories of the text.

The economic community that succeeded the USSR on October 18 thus appeared to be a demonstration of good will, quite unlike a serious framework for organizing an economy based on lasting commitments. At this time, one could glimpse that beyond the plan, from which all concrete problems had been omitted or deferred to the signing of later "special agreements," the good will of the signatories had been primarily directed toward the industrialized countries and probably represented a wish to convince them that the former USSR remained an economic partner with which they could do business.

As watered down as this treaty of economic union was, it soon appeared too constrictive for the states concerned. Those who had not agreed to sign it issued declarations underscoring their refusal to join the community. And by adopting, on November 16, a plan of its own for immediate economic transformation, Russia, whose support would have given meaning to this last attempt to save the appearance of a common plan, dealt it a fatal blow.

New Plan, New Failure

On October 21, the first session of the postcoup variant of the remodeled Supreme Soviet was to assess the Soviet breakup. Despite the signing of the economic treaty three days before, only seven republics attended the session (there had been eight seventy-two hours earlier). As in Agatha Christie's well-known mystery *Ten Little Indians,* the former family of Soviet peoples lost a member with each passing day. And this handful even had to put off a number of discussions in the hope of seeing some representatives of Ukraine show up. But none did, except as observers, and this much-reduced assembly of republics listened with a certain skepticism to Gorbachev's announcement that the debate on the union treaty would be resumed on the basis of a new plan. The latest plan—containing twenty-three articles—represented an odd compromise between Gorbachev's still-present unionist hopes and the dismantlement of the center by the authorities of the Republic of Russia. Despite this plan's very general clauses and principles—adherence to democracy, human rights, the UN charter, and so forth—articles one to eight defined the basis of the union by largely restoring the federation on the earlier model—with a common constitution, a president elected by universal suffrage, a large area of common interest. The Union of Sovereign States (with unchanged initials) wished to be a sovereign state, subject to international law, and the successor to the old USSR. These claims did not satisfy most of the states called on to join it.

At the October 21 session of the overhauled Supreme Soviet, the assembly, which Gorbachev was seeking to involve in the plan by calling it a constituent assembly, proved indifferent and at times hostile to his appeals. It began by shooting down the idea of a constitution peculiar to the union. In the last analysis, it was the State Council, an institution from the interim period in which the president of the former USSR coruled on an equal basis with the presidents of all the republics, that on November 14 took a position on this alarming specter that the union treaty had become. Certainly, the members of this council—they were only seven of them—said they agreed on forming a union, but they immediately reduced it to its simplest expression—more an agreement than a state. And that posed a serious political problem that the international community still wished to ignore but that the politicians of the former USSR boldly confronted—the representation of the future union (if it came into existence) and the republics in international organizations.

Since the coup, the representatives of the USSR—Boris Pankin, minister for foreign affairs up to November 20, 1991, and his vice minister, Vladimir Petrovsky—endeavored to persuade the international authorities that in any event the future union must keep its seat on the UN Security Council. Boris Pankin linked what he held to be a permanent right to a council seat to the nuclear power of what had been the USSR. Whatever its future name, the former USSR was still a nuclear superpower and must consequently remain an international superpower. Pankin's argument was two-pronged. On the one hand, he was playing on the world's anxiety about a still-intact nuclear capacity liable to possible chaos, in order to demand that by leaving its international status unchanged, the international community would support and even preserve what remained of the USSR. On the other hand, he clearly intended—as did Gorbachev—to use this international status tied to the Soviet seat on the Security Council to show the republics that the USSR still existed and that they could not bolt from the union.

It is doubtful that the republics were willing to maintain this unity in international affairs. The three Baltic nations had won representation at the UN; for the time being, the three republics of the Caucasus—Armenia, Georgia, and Azerbaijan—had requested but not been given the same treatment. It is doubtful, however, that the UN could long ignore the states emerging from the breakup of the USSR (as from that of Yugoslavia). The formerly colonized African countries that gained independence and were received into the international community were a reference point for many in the former USSR. And Armenia and Georgia, sites of ancient civilizations, could not be treated differently from Benin.

Already, the states who grappled with the complications of the UN, or who feared struggling with them, were seeking to gain time by creating a de facto situation by roundabout means. Thus, Uzbekistan asked Ukraine, a member of the UN since 1945, to assume the task of representing it in the international authorities. The Uzbek leaders were reasoning that sovereignty implies the possibility of choosing the channel of representation that is most consistent with national interests. But it was the Republic of Russia that, implicitly, most seriously challenged the Soviet state's future capacity for international representation. By proposing to the republic's Supreme Soviet on October 28 to dismantle local authorities, notably of the ministry of foreign affairs, and make them an organization that simply coordinated the foreign policies of the republics, Boris Yeltsin dealt a blow both to the Soviet ministry and to the distinctive role of the USSR, a blow whose impact has yet to become fully apparent. The ministry, the ex-MID (*Ministerstvo innostronnykh del*—ministry of foreign affairs) had, moreover, been rechristened *Ministerstvo vnechnikh sviazei* (ministry for foreign relations) and was already subject to a policy of insidious "invasion" of authority by certain republics. Besides its reduction in personnel and spheres of activity, unilaterally decided by Boris Yeltsin, the ministry found itself in competition with its counterpart in the Republic of Russia and its head, Alexis Kozyrev. Already, certain posts had been filled by Boris Yeltsin's nominees. It may have been to retard this process of removing the remaining last source of central authority—the one it enjoyed in the international arena—that in mid-November Mikhail Gorbachev decided to get rid of Boris Pankin and reinstate Eduard Shevardnadze (who had quit this post less than a year previously) as head diplomat. Thus in less than twelve months—December 1990 to November 1991—the USSR had four ministers for foreign affairs: Shevardnadze, who resigned; Alexandr Bessmertnykh, who was put out of business by the fiasco of the coup; Boris Pankin, who was appointed as a reward for deft actions during the coup but who resigned because of the nonexistence of Soviet diplomacy; then Shevardnadze again, recalled to his post like a doctor summoned to the bedside of a critically ill patient.

Can Shevardnadze's return to business straighten out this badly compromised situation? Certainly, his personality reassures to some degree Gorbachev's foreign associates, who were distressed at the lack of a familiar and reliable interlocutor. In less than a year, however, the situation in the USSR has changed so much that his ability to restore a viable diplomacy may be doubtful. He himself has felt this, since his first ministerial action back in his post was to start touring the republics to convince them to allow him to represent them. At this point, two questions arise: Can Eduard Shevarnadze move beyond the role of representative—or a simple coordinator—of the foreign policies of the large republics? Does Mikhail Gorbachev's loss of both domestic and international (as the start of the Middle East peace talks in Madrid made cruelly evident) authority also run the risk of hurting Shevardnadze's own image? Everything suggests that Shevardnadze could go down with Gorbachev in a fall that would be hard to stop. History is generally not inclined to favor attempts at restoration.

Russia and Ukraine: The True Debate

The question of the union, which continues to agitate what was the center, has for the most part been overtaken by two partially connected problems: first, that of the parallel emancipation of *both* Russia and Ukraine, and next, that of the relations between them and, more generally, Russia's relations to the other republics.

Ukraine, long cautious in its aspirations for independence, suddenly and quickly stated some extreme demands. For a long time, the Soviet government counted on the soothing certitude that the complexity of this republic, its links to the union's authority—the Ukrainians had provided it with a considerable proportion of its bureaucrats and officers, and they also have a seat at the UN, a privilege shared with Belarus but not with Russia—and the presence of a large Slavic community would neutralize many of the temptations of independence. Certainly, its division into three distinct regions—western Ukraine, belatedly Sovietized and fervently Catholic even though it was forcibly annexed to Soviet orthodoxy after 1945; eastern Ukraine, widely subject to the pressures of Russification; and the industrial Ukraine, populated by Russians—was long a source of weakness for the nationalist movements. But recent events had somewhat modified these divisions. Western Ukraine spearheaded the nationalist movement. The referendum of March 17, 1991, and even more, the coup, encouraged the rapprochement between two Ukraines, eastern and western, and the radicalization of their demands. To avoid being overwhelmed, Leonid Kravchuk, the president of the Ukrainian Supreme Soviet—a product of the Communist party who refused to leave it—made the nationalist demands his own: the setting up of a totally responsible state with complete economic and military independence. All or nearly all the Ukrainian political groups, from the Communist party to the ardent anticommunists, but also RUKH, the popular front, agreed on a platform stipulating

- That Ukraine cannot sign any union treaty without previously taking a stand on independence and equipping itself with a constitution;
- That in any event—with or without a union—it had to achieve complete economic independence, notably by breaking with a common monetary system; the Ukraine vowed that it would not compromise on the principles of printing a specifically Ukrainian currency, the control of the money supply, and absolute equality between the Ukraine central bank and a hypothetical central bank of a no less hypothetical union;
- That for the future, the Ukraine has refused to contribute to a common army; the minimal demand of its leaders is to keep all its recruits on the national territory and the Ukrainian government's control over the Soviet troops stationed on its soil. Even more vehemently, however, the leaders of the Ukraine have stated their preference for a national army of between 300,000 and 400,000 troops, and they quite correctly recall that the number of professional officers of Ukrainian nationality serving in the Soviet army would be perfectly adequate for the staffing of an army of this size.

To the military issue must be added the nuclear problem. A considerable proportion of the Soviet strategic and also tactical nuclear potential is located in Ukraine. Until recently—in fact, up to the coup—Ukraine's position on this topic was very simple: Ukraine demanded the withdrawal of all nuclear weaponry from its territory, which it wished in the future to be nuclear-free. Since the coup, this position has become more nuanced. While retaining the goal of eventual denuclearization, Ukraine wishes to be a party to this process and refuses to lose a means of persuasion too soon. This development allowed Mikhail Gorbachev to use, perhaps more effectively than in the past, the argument that a "center" was needed to prevent a nuclear dispersal feared by the whole world. A Ukrainian state with a population of some 50 million people, an army of 400,000, and an arsenal of nuclear weaponry would indisputably upset Europe's strategic balance.

Although the coup stepped up a widespread demand for Ukraine's independence, it is noteworthy that even before August 19, Leonid Kravchuk gave all his political positions a virulent nationalist twist. Addressing the Ukrainians in celebration of the first anniversary of the declaration of sovereignty, described by the Ukrainian press as the "day of independence" (*Pravda Ukrainy* July 14, 1991), Kravchuk painted a dramatic picture of the price Ukraine had paid for its attachment to Russia. Concerning the prerevolutionary past he said, "For us, as for other people, moreover, the time [that of union with Russia] was not a time for developing our national culture, material and spiritual." Condemning this period and the tragic consequences for Ukraine of its inclusion in the USSR, Kravchuk in passing hailed the memory of all the Ukrainians who had attempted to advance the cause of a national state. Thus the communist Kravchuk found himself in the same camp as the fiercest anticommunists in Ukraine.

This psychological evolution was confirmed by the referendum on Ukraine's independence on December 1, 1991, which accompanied the election by universal suffrage of the republic's president. Nearly 80 percent of the electorate turned out, which meant that all or nearly all the Ukrainian and Russian voters massively supported independence. Mikhail Gorbachev's entourage had long considered the talk of independence as being simply part of the electoral campaign but without further consequences. In any case, the remaining leaders of the center still wished to believe that, after the referendum, Kravchuk would have no reason to delay signing the union treaty. That meant ignoring the whole course of Ukrainian policy, which for a year had formed a network of relations between republics, shutting out the center from the future organization. The Russia-Ukraine treaty, Ukraine's accords with Belarus and various Central Asian republics, and also, outside the former Soviet

Union, the establishment of contractual ties with Germany and Hungary were all signs of a desire to deny the union international spheres of jurisdiction.

Ukraine still has to deal with the areas populated by Russians. The republic has some 11 million Russians within its borders. What would their fate be under conditions of separation? For the most part, these Russians have no intention of becoming culturally Ukrainian. In Ukraine as elsewhere, the trend is to impose the use of the republic's official language on all inhabitants. Nor is it an enticing prospect to become resident aliens. In the present chaos, however, Russia cannot envision the return of 11 million refugees. The threat of the Russians' seceding from Ukraine now weighs heavily on the dream of independence and on the relations between Russia and Ukraine.

The thundering entry of Russia into this race for sovereignty is another decisive factor in current developments. Here, too, 1991 was marked by the overturning of every certitude.

Just after the coup, Boris Yeltsin proclaimed, "Russia has returned!" and it is true that his political ascendancy is connected with his desire to salvage Russia. The brilliant victory of his election to the Russian presidency on June 12 was dominated by his position on Russia. Waving the tricolored flag in meetings, fortified by his break with the Party and his internationalism, Yeltsin was truly a national candidate. Therefore, his success indicated that Russia was differentiating itself from the Soviet Union. If one went beyond the literal words of his speech to examine its content, his position on the treaty and the preservation of a union was clear. No doubt Yeltsin said that he wanted to keep a common structure, but he also said that this structure did not imply any supremacy of the central authority over the Russian ballot boxes or any federally imposed system or central jurisdiction over the natural resources and enterprises scattered across Russian land.

After the coup, Yeltsin's carefully phrased remarks, designed to veil the growing nationalism of his compatriots, were softened as the central government—and Gorbachev personally—emerged from the ordeal in such a weakened condition. Fortified with the sense of adding a new legitimacy to the voting, that of the savior of democracy, Boris Yeltsin quickly went about endowing all the authorities of the central political system with Russian political structures. This was particularly noteworthy in the area of foreign policy and disarmament, which had hitherto been the preserve of the central government. The tour of world capitals by the

Russian minister of foreign affairs, and Yeltsin's September 7 announcement of an eastern policy—"We must extend our foreign policy to Asia"—were signs that he henceforth intended to conduct his own foreign policy. In September, by bidding up Gorbachev's disarmament proposals (a 50 percent cut in the nuclear arsenal of the Big Two), Yeltsin coolly encroached on a reserved area.

His November 1991 program for radical and immediate reform was also an obvious negation of the union. By decree, Yeltsin "Russianized" the federal ministry of finances, decided to go all the way with privatization and the changeover to a market economy, announced the semiconvertibility of the ruble, freed foreign trade, and placed under Russian control all natural resources and the share of the financial system that Russia could demand. The whole of this ambitious program, conceived for Russia alone, reduced the union to an empty shell and broadcast to the other republics that Russia was ready and able to go it alone even if they did not acquiesce. When in late November Russia took over the obligations of a Soviet state in the process of suspending payment, the replacement of the former USSR by Russia over the state institutions and their personnel was complete.

The signals thus sent were primarily aimed at Ukraine, which was rightly frightened by this Russian rebirth. Here the problem of Russian-Ukrainian relations, which had deteriorated since August in a nearly unstoppable series of events, resurfaced.

Unquestionably, the coup intensified the Ukrainians' desire for independence. Faced with the prospect of a union without Ukraine, the Russian leaders reacted with great vigor and extreme rashness. Boris Yeltsin produced a real shock in Ukraine by proclaiming that Russia would oppose an independent Ukraine on the issue of its frontiers, raising the idea of a possible secession of the areas of central Ukraine populated by Russians and, above all, by posing the problem of the Crimea.

For the Russians, Crimea, a Tatar country annexed by Catherine II in 1793, was both a symbol of their definitive victory over the Tatars, who had dominated it for nearly three centuries, and an outpost on the Black Sea. Although in 1954 Stalin's successors annexed the peninsula to Ukraine, they could not then foresee that Ukraine would one day promote the idea of seceding. And because the Crimea was now emptied of Tatars whom Stalin had deported to Central Asia for "collaboration" and who were not allowed to return, the central government found it convenient to assign to the Ukrainians the guardianship of the peninsula and the job of eradicating any trace of the Tatar civilization. Nearly forty years sufficed to convince the Ukrainians that Crimea belonged to them for good, but forty years certainly did not suffice for them to accept its loss to Russia. All the more so as Boris Yeltsin's declaration was reinforced by one of his closest collaborators, Gennadi Burbulis—"Russia is the lone republic that can and must become the legal inheritor of all the structures of the Soviet Union" (*Izvestia,* October 2, 1991)—and by the proposals of Alexandr Solzhenitsyn, hitherto sparing in his comments on the situation in the former USSR, who suddenly spoke out on the Ukrainian referendum of December 1. Opposed to Ukraine's secession and supporting a union of the three Slavic states—an idea that he had discussed at length in 1990—Solzhenitsyn suggested that the results of the approaching referendum should be taken into account region by region, and not globally, so that the population of each region (and hence the Russians) could decide their future by themselves ("Declaration on the Referendum of December 1, 1991," *Trud,* October 8, 1991).

This referendum on the Crimea, where the Ukrainians are in the minority and Russians in the majority (out of the 2.5 million inhabitants of the republic, close to 70 percent, that is, 1,600,000, are Russians, 700,000 Ukrainians, and 100,000 Tatars recently authorized to return) would have the effect of allowing the Russians to force the Crimea's annexation to Russia. Already, the referendum of January 20, 1991 to define the status of the Crimea ended in 93 percent of the voters favoring the autonomous region's conversion to a full republic and a direct member of the USSR, and hence separate from Ukraine. The referendum had no practical consequences but was designed to make the Ukrainians think.

This is a source of conflict that Boris Yeltsin has, consciously or not, made clear. His brusque statement immediately convinced the Ukrainians that the danger for them came both from the former center or a possible union and from a renascent Russia. The Russian-Ukrainian cooperation symbolized by the bilateral treaty of November 1990 suddenly gave way to feelings of distrust and even overt hostility. Ukraine slapped an embargo on its grain exports to Russia and ever more strenuously asserted its intention to arrange for its own defense. Yeltsin was far from isolated in his anti-Ukrainian campaign. High Russian officials like Gavril Popov and Anatoly Sobchak shared his conviction that Ukraine could not secede and liquidate a union that was more than three centuries old or it would have to pay the price for it. Although some of the most liberal Russian democrats, such as Yuri Afanasiev, waxed indignant that their friends scoffed at the right to self-determination, they were in the minority.

People close to Boris Yeltsin declared that Ukraine is not Central Asia, which is alien to the Slavic community and thereby more entitled to secede from it.

Meanwhile, the "battle of the Crimea" also started up in the disputed peninsula. The recently formed Movement for the Republic of Crimea, which was close to the local Communist party, held numerous street demonstrations and hunger strikes in support of the idea that the 1954 decision taking the Crimea away from Russia was illegal and so must be rescinded. It was easy for the Ukrainians to answer—and they were not wrong—that this was a provocation and did not reflect the mood of the majority who did not care, as public opinion polls confirmed, about a change in their region's status, except to become a rightful republic, independent of Russia.

Ukraine was not the only republic worried about the rise of Russia. Kazakhstan, where nearly 7 million Russians lived, also heard Boris Yeltsin's warning to the aspirants for independence and his insistence on the frontiers and Russian-populated areas that he could claim. If all the former components of the union could be separated, Russia could incorporate the prosperous northern part of Kazakhstan, with its concentration of Russians. This implicit threat helps explain why Nursultan Nazarbayev, the president of Kazakhstan, advocated that the union be maintained in any form whatever and worked so energetically to save some elements of the former USSR.

What is more, Ukraine and Kazakhstan were in similar positions regarding nuclear weaponry. Possessors of a share of the Soviet strategic potential, these two republics both wished for total denuclearization in the future. But they worried about Russia's demands for the return to Russian territory of all nuclear weapons stationed in their republics and in Belarus. General Lobov, the Russian chief of staff, strongly supported the idea—notably in an interview in *Krasnaia Zvezda* (October 20, 1991)—that the USSR's whole nuclear arsenal had to be transferred to Russia. These declarations fostered a mood of panic in Ukraine, reinforced by an article in the *Moskovskie Novosti* (October 20, 1991) stating that Yeltsin had seriously considered the possibility of a preventive nuclear attack against Ukraine and dropped this plan only because he judged it impossible for purely technical reasons. As implausible as they were, these rumors testify to the degree of tension in the relations between the Russians and the Ukrainians on the eve of the referendum of December 1, 1991.

The same was true of Boris Yeltsin's disarmament proposals—an attempt to outmatch those of Gorbachev—in which the USSR's nuclear republics understood that top priority would be given to destroying nuclear weapons located outside Russian territory. These republics could not tolerate being denuclearized next door to a Russia that still controlled the strategic force of the former USSR. This explains the seemingly contradictory resolution by the Ukrainian Supreme Soviet on October 24, 1991: in this resolution, the legislature affirmed both its allegiance to the principle of denuclearization, its desire to adhere to the nuclear nonproliferation treaty, and also, for a certain transition period, its intention to keep nuclear arms on Ukrainian soil and retain control over them conjointly with the Soviet (and not the Russian) government.

Belorussia, from which Russia also wanted to repatriate nuclear arms, was in turn undergoing two developments. The assertion of a national, as opposed to a Soviet, identity was added to a growing desire to dissociate itself from Russia as well. On September 19, 1991, the republic's Supreme Soviet voted for a change of name, replacing Belorussia with Belarus and the starred red flag with the red and white national flag and the old arms as symbols of the nation. Above all, the vote affirmed that Belarus was to have genuine sovereignty in relation to *both* the union and Russia.

Considering the waves of fear caused by the "Russian plan" indisputably spearheaded by Boris Yeltsin, we can wonder what the center of attraction and equilibrium of the union would be. The federation advocated by Gorbachev, drained of most of its prerogatives by the leaders of the republics during negotiations about its future, was no doubt seen as a necessary structure for limiting Russian designs and forcing Russia not to keep its immense natural resources for its benefit alone. Could the other republics, especially the poorest ones, accept being deprived of the gold, diamonds, and oil that assured Russia of its future? Hadn't Boris Yeltsin taken care to place them solely under Russian authority? Certainly, these resources are concentrated in Russia, but during the time of the empire and in the USSR, the common good of the peoples required that these resources be shared by all. In 1991, the republics found out that Russia wished to be sole owner of these resources, which would increase its power while hindering economic recovery and progress in the states without them. The republics thus vacillated between their dreams of independence and the desire, while preserving a minimum of shared governing structures, to increase their chances for a reconstruction that very few of them could accomplish alone.

For the supporters of the union, the main difficulty in solving this problem lay in the gulf separating the republics that could survive on their own and those for whom that was impossible. Russia and Ukraine can survive, and, for this reason, Ukraine is little tempted by the union and envisions a future of independence and rapprochement with the countries of Central Europe and even the European Economic Community. The Ukrainians ask, "Are we, with 50 million citizens, less worthy than Hungary or Czechoslovakia of being considered a large European country?" And if Ukraine reached an extreme solution—separation from its former allies—then could it be denied that, without Ukraine, Russia could not live on in a group of countries inherited from the union? This group—including nearly 300 million people, of whom 220 million are Slavs numerically dominated by the large Russian population of 150 million—make Russia the arbiter and even the real heir of the former USSR.

In the future community, though, Russia (for in the event of Ukraine's secession, Belarus had every chance of going the same route) will have to confront the large (nearly 60 million) and demographically dynamic Muslim bloc, and it is likely that Russia would judge this confrontation too deleterious—for demographic reasons (a stagnating Russia and a growing Muslim periphery); for economic reasons (Russia would have to bear the full weight of the Muslim periphery with its worsening underdevelopment while Russia itself is finding it hard to specify the conditions for its own recovery); for psychological reasons (the Russians, who feel they have been bled dry for the benefit of the periphery, will no longer make the slightest effort on its behalf); and finally and most important, for political reasons (Russia wishes to be European, and the Muslim periphery is rapidly "re-Islamizing" itself, underscoring its distinctiveness from and opposition to Russia). Without the support of the Ukrainians and Belarussians, the encounter between Russians and Muslims would have the huge drawback for Russia of drawing it into the "oriental" and even Asiatic destiny that throughout its history it has wished to escape. What is more, Russia is a prey to internal upheavals similar to those that wreaked havoc in the USSR, and this also forces it to give first priority to its own cohesion and interests.

Internal Agitation in the Republics

While the USSR existed and figured as the oppressor in the eyes of all the nationalist movements, no one imagined that the republics, so determined to assert their rights, could in turn become the victims of internal nationalistic demands. Following the logic of the conflicts between the nation-states, the ongoing conflict in Karabakh, pitting Azerbaijan and Armenia against each other, continues to produce many deaths, the number of which is kept secret. For a while, it has somewhat overshadowed the growing importance of conflicts within the states, which as the months go by increasingly reduces their chances for nonviolent transformation.

The chief victim of these conflicts is Georgia. In less than a year, this multiethnic republic saw the rise of nationalist and territorial conflicts within its borders. Three very different cases created an uproar that gradually drove Gamsakhurdia's government to turn radical and meet all demands with an extremely authoritarian attitude. To understand these cases, we should recall the major stages in the political changes occurring in the republic. In October 1990, the local elections (belated in relation to the Soviet agenda) were marked by the resounding success of the Free Georgia Round Table conducted by Zviad Gamsakhurdia, who became president of a Supreme Soviet dominated by his list of candidates. In the national referendum of March 31, 1991, 98 percent of the voters favored "the restoration of the independence of Georgia based on the declaration of independence of May 26, 1918. The massive vote was all the more remarkable in that 90 percent of the registered voters went to the polls. On April 19, Gamsakhurdia got the 227 members of the Supreme Soviet unanimously to adopt the declaration of independence. Although it is hard to doubt the sincerity of the deputies' votes—like that of the voters in the elections and referendum—the "Gamsakhurdia method" seemed to contravene democratic principles. First, without prior debate or real voting, he had his parliament vote by simple acclamations that gave the impression of being directed at least as much at him as to the topic being voted on. Above all, Gamsakhurdia excluded a part of the population, the Ossetians, from joining in the game of self-determination that he had just won for the Georgians. Georgia has to deal with three highly problematic territories—Abkhazia, Ossetia, and Adzharia. While Abkhazia owes its status as an autonomous republic to the presence of an ethnic minority—

the non-Georgian Abkhazians—and Ossetia was set up as an autonomous region for the same reasons, the Adzhari republic was founded on religious and not ethnic criteria, as the Soviet system wished. There are, strictly speaking, no Adzharis but Georgians who converted to Islam between the sixteenth and seventeenth centuries, when this part of Georgia was under Turkish rule. This region was retaken in 1878 by Russia, then returned to Turkey in 1918, and finally retaken by the Soviet power under the friendship treaty of 1921, the product of Lenin's support of Kemalist Turkey. Lenin placed Adzharia under Georgia's jurisdiction but did so on one condition—that its profoundly Islamized population acquire an autonomous status.

The rise of Islamic feeling in most of the USSR's republics did not bypass the population of Adzharia. Long indifferent to Islam, these ethnic Georgians have in recent years shown ever more clearly that they feel they are Muslims and have adopted traditions alien to Georgia—Islamic religious practices, polygamy, the wearing of veils, and so on. More serious, they have turned to neighboring Turkey, which once Islamized them, and demanded the right to be annexed to it. In the elections of October 1990, the Adzhari region massively voted against Gamsakhurdia's coalition, which was seen as a nationalist party hostile to any difference and hence to the one that Islam represents. Just after the elections, Gamsakhurdia's response was to propose to terminate the region's autonomous status, arousing fear and rebelliousness among its inhabitants.

The Adzharians' discontent, which fueled their yearning for separatism, thus linked up with the Ossetian movement. Since the 1920s, the Ossetians have had an autonomous region—southern Ossetia—where 65,000 of them live, that is, two-thirds of the total population. There are more Ossetians (335,000) outside Georgia, in the autonomous republic of Northern Ossetia, which is part of Russia. Two claims are in conflict on this issue. For the Georgians, the Ossetians of Georgia are simply migrants—aliens with no historic right to the land on which they live and hence with no political rights. Even though the southern Ossetians, like the Georgians, are Christians (the northern Ossetians are Muslims), the Georgians would see only a gain in the Ossetians' returning to the north and thus vacating a disputed territory. Although the Ossetians also think that their interests would best be served by rejoining the Ossetians to the north, they judge that this reunification of a people would also imply the reunification of its territory. Their separatism thus impugns Georgia's territorial integrity. In 1990, the conflict quickly became violent. The

Georgian government stepped up restrictions on the rights of minorities, notably by imposing the universal use of the Georgian language throughout public life. "Violation of the Constitution of the USSR," protested the local authorities, who appealed to Moscow to arbitrate the conflict. In September 1990, while acts of violence and the number of deaths climbed—troops from the Georgian ministry of the interior crisscrossed the region—Ossetia proclaimed itself a sovereign republic and requested its immediate incorporation into the USSR. The Georgian response was brutal: the dispatch of additional troops to put down any public demonstration and, above all, the revocation of the region's autonomous status by a legislative vote on December 11, 1990. For Gamsakhurdia, the Ossetians were simply pawns in the Soviet game to quash Georgia's desire for independence. Therefore, he had to deprive them of any means for political expression. Georgian independence further widened the gulf between Ossetians and Georgians, for Gamsakhurdia had proclaimed both his desire to humor the Abkhazians, whose autonomous status he had promised to protect, as well as the Adzharis, but he proved unyielding toward the Ossetians. The latter had enough small arms to turn the local situation into permanent guerrilla warfare. Armed groups of Georgians, convinced they are defending the republic's independence and territorial integrity, go on punitive expeditions to the Ossetian villages. Hence, the situation has few chances of calming down.

Joining the problem posed by nationalities with a special status are the national minorities who have been living in Georgia for centuries. Peacefully coexisting with the Georgians, but hungering for independence and exhibiting a heightened nationalism of their own, these minorities wonder about their future. Here, the main problem is that of the Azeris. The total Georgian population of 5,395,840 includes 307,500 Azeris, but their much faster demographic growth than that of the Georgians (a 10 percent increase in ten years for the Georgians, 20 percent for the Azeris) make the Georgians afraid of being eventually outnumbered by rapidly growing minorities, and hence they display great aggressiveness toward them. The Azeris and a few other small groups, like the Lezches, are accused of appropriating better lands and illegally building on plots that rightfully belong to Georgians. Threatened with being deprived of their property, not benefiting from citizenship in an independent Georgia, which is reserved for people whose ancestors lived there before the annexation of 1901, and thereby deprived of their rights, the Azeris think of fleeing a state that rejects them. Meanwhile, in the settlements where they live, guns are coming out of their hiding places.

Beset by minorities who consider independent Georgia a dangerous nationalistic state and wish to split off from it, the government of Zviad Gamsakhurdia, who attributes minorities' anxieties to the Soviet—or Russian—"conspiracy," is locked into growing intolerance and violence. Far from leading to democracy, Georgia's independence has produced authoritarianism and the rejection of all the minorities that had shared the Georgian fate for centuries. This development was correctly diagnosed by Andrei Sakharov, who, before his death, stated that Georgia, like Russia, was a miniempire that was condemned, like Russia, to modernize itself and accept the right of peoples to self-determination and equitable treatment.

After Georgia's victorious combat with the communist Goliath, the nationalist issue that destroyed the Soviet power now eats away at the republic like leprosy.

Equally threatened by internal disintegration is Russia, the other "empire" in Sakharov's eyes. The censuses of 1979 and 1989 counted ninety-seven ethnic groups on Russian soil. Some of these groups, of course, amount to just a few hundred resident aliens—French, Americans, and Germans. In other cases, however—Izhors, Karaims, Krimchaks—fewer than a thousand people form an ethic group that has lived for centuries on Russian soil. The grouping of the largest national communities into (sixteen) autonomous republics, regions, and districts was not enough to satisfy their desires for full recognition of their national identities. Among the mass of people settled in Russia, some have already made demands that the Russian republic does not wish to meet. The most urgent problems are posed by the Tatars, the Chechens, and the Ingushes, of course, but also the Buriats, the Yakuts, and the Germans.

The Tatars (this name excludes the Tatars of Crimea, only a hundred thousand of whom were allowed to return to the homeland from which they were deported in 1943) have an autonomous republic where, with 1,765,400 people, they represent half the total population of 3,641,000, followed by the nearly as numerous Russians (1,575,000).

On August 30, 1990, the Tatar Republic proclaimed its sovereignty, which, in the great wave of such declarations, was nothing out of the ordinary. What was unusual was the radicalization of the Tatar positions. Regarding the content of this sovereignty, the Tatars clearly expressed their desire to enter the Soviet federation as a fully fledged republic with no connection to Russia. They were also among the first to assert their full property rights over their national resources—particularly oil (this claim was immediately echoed by the Yakuts, whose subsoil is one of Russia's main sources of gold). The return to the old name of Tatarstan is no less revelatory of the underlying Tatar ambitions, since the name Tatarstan (land of the Tatars) is fraught with great historical and ideological content. As the center of Islamic civilization and unification for Russia's Muslims, the Tatar country has exerted its influence over all of Central Asia. Stalin's energies were mobilized for a long time in the struggle against "Tatarstan's hegemonic aspirations." As a result of the energetic resistance of the Russian central government to Tatar separatism, the authorities of this republic broke away from it and on October 8, 1991, decreed the formation of a national guard—or national military units—based on voluntary service. Many volunteers came to Kazan to put themselves at the service of the local authorities. The purpose of the army's nationalization was to defend the new republic against any steps Russia might take to destroy Tatarstan's sovereignty and to ensure the handing over to the republic's authorities of all Russian or Soviet enterprises on the national soil. After these measures, the Tatars are dreaming of total independence, reasoning that a national referendum would free them to follow the example of the Baltic countries.

Just as Yeltsin stripped the USSR of its authority and resources in order to give them to Russia, the Tatars similarly intend to strip Russia of its authority and the resources that they take to be their own property.

Although they entered more belatedly into the nationalist disputes, the Chechens and the Ingushes are no less a source of anxiety for Russia. These two peoples wield considerable influence. Close to a million Chechens and 238,000 Ingushes live in the former USSR. Deported by Stalin in 1943 and deprived of their nationhood, these people were later "rehabilitated" by Khrushchev after the Twentieth Congress and resettled in the densely concentrated Chechen-Ingush Autonomous Republic, a part of Russia. Today 735,000 Chechens and 164,000 Ingushes form the main body of the republic's population of 1,270,000, supplemented by close to 300,000 Russians.

Chechens and Ingushes retain bitter memories of the treatment inflicted on them. (Documents recently made public from the Soviet ministry of the interior have yielded more precise and tragic estimates of the cost of these deportations.) It thus appears that between the time they were ejected from their land and the time they arrived at their places of exile, they recorded human losses amounting to a quarter of their population, in addition to their losses during their exile, where conditions were just as frightful. The Chechens and Ingushes have forgotten nothing and rightfully speak of genocide. And, at a time of accountability and triumphant nationalism, they in turn have risen up. Well armed—as who in the Caucasus is not?—they are engaged in a struggle for sovereignty that they believe in all likelihood will lead to total independence.

The national problem in Chechen-Ingushetia is aggravated by an additional issue. The anticipated break with Russia has spurred a conflict between the two main ethnic groups. Although both agree in forming armed units to resist pressures from Moscow, the Chechens have proceeded to make themselves politically dominant in the common state. With only the Chechens taking part, the elections of October 27, 1991—which the Russian Supreme Soviet declared illegal—gave birth to the Chechen Republic, with its capital at Grozny. An appeal was then launched "to their [Ingush] blood brothers," inviting them to make themselves at home in Chechen. For Russia, although it was assured of allies—the Ingushes who were ousted from power—this internal division is in no way reassuring. The two groups clashing and firing at each other instead of at Moscow is an additional cause of anarchy in a Russia with no need of it. And no Russian can forget that of all the wars of the imperial era, the hardest, longest, and costliest in human life was precisely that one that ravaged neighboring Daghestan, under the leadership of the invincible imam Shamil. His ghost now hovers over the conflagration that is consuming Chechen-Ingushetia and rightly worrying the Russian authorities.

Another problem must soon be resolved, a nonviolent one, certainly, but still very important for Russia's international position—that of the Germans. The Germans of the Volga, also deported by Stalin, number 2 million. Although the Soviet government could long ignore them, that is no longer possible, if only because Germany, which Russia is counting on for its recovery, has made this a condition for its aid. And helped by "the mood of the times," these Germans are demanding that they be given their own territory. In 1991, they held a national congress to symbolize their future territorial and political reunion. And Boris Yeltsin made himself the champion of a serious territorial solution. Making a state visit to Germany in 1991, he committed himself to the reinstatement of the Germans' political status without delay. The site of the future German republic has already been determined—in the region of Saratov. But here again, national sensitivities are hard to overcome. The region's inhabitants—mostly Russians—are opposed to their eviction. And their hostility fuels the Russian extremist movement that attacks Boris Yeltsin's policy as "dismembering" Russia at the instigation of indifferent collaborators or strangers to the Russian cause. Yeltsin's chief adviser in national issues is a Chechen, Ruslan Khasbulatov. For the extremists, who are still marginal but who have understood what an opportunity the nationalist ferment in Russia presents for them, it is easy to turn this situation to account and make headway by telling the Russians, already the invalids of the empire, that they are also threatened with eviction from their own country.

Islam and the Pan-Islamic Temptation

During the early years of Gorbachevian change, a portion of the USSR's Muslim area seemed destined for stability rather than the upheavals increasingly affecting most of the country. At one time, it could be believed that although the Soviet empire was threatened with dislocations in the west—the Baltic countries and Ukraine—it could count on the fealty of the Muslim periphery. No doubt this conclusion was based on an ignorance about the agitation and climate of violence descending on the Muslim Caucasus. This conclusion also ignored the growing number of Tatar dissidents. Finally, it did not take account of one decisive factor: the belated awareness in the Central Asian republics of the reality of their situation—real and deteriorating underdevelopment; alarming unemployment (close to 30 percent of the population); an ecological disaster that made any hope of recovery in the region impossible; a real collapse in sanitary conditions; and a complete cultural inadaptability to any policy of progress. At the same time that the populations were taking stock of this situation, they realized that their survival would depend on the transfers of resources within the USSR. Under these conditions, how could they defend independence claims?

There were three consequences of this sudden awareness. First, the people of the region wanted to preserve the teams in power combining communist legality with mafia law because these teams seemed to guarantee the maintenance of the existing order. To preserve the USSR or die of hunger: this was the alternative at the end of the 1980s. But this resignation to the maintenance of the USSR was offset by an extraordinary bitterness toward the Soviet state, communism, and the Russians, attacked en masse for the backwardness of Central Asia and suspected—with reason—of wishing to get rid of it to better ensure the progress and political survival of the western part of the empire. The final and not least consequence was withdrawal into Islam to find cohesion and a way of responding to the challenge of a possible abandonment by the USSR. During the last three decades, Islam—often in its fundamentalist form—has become a key element in the refashioning of the societal landscape. In Tadzhikistan, Uzbekistan, and also the belatedly Islamized Turkmenistan, the rise of Islam has been incontrovertible—and with it, the rise of pan-Islamism. The Muslims of the USSR are trying to link up with the Muslim countries that are powerful or could be so—chief among them Iran and Saudi Arabia. The growing nationalist movements, often with Muslim overtones, represent, like Islam itself, an outlet for peoples "condemned" to save the USSR in order to save themselves. An undeniable sign of this increasing unification of Central Asia, and even the Caucasus, through Islam was the founding in June 1990 of the Islamic Renaissance Party in Astrakhan; this party has since expanded to the whole of the Caucasus and Central Asia, beginning with Tadzhikistan and Uzbekistan. The *Literaturnaia Rossiia* of March 8, 1991, reported an important investigation of this party that claims to be modernizing and Muslim (perhaps a current variant of turn-of-the-century Dzhadidism, to which it sometimes makes reference) and highlights its rapid spread and organization. Fundamentalist Islam, modernizing Islam, and nationalist parties like the Birlik in Uzbekistan: one can only note that Muslim society is organizing by drawing on its distinctive traditions.

Moreover, in this profound transformation of societal and political aspirations, two republics are vying to play a central role in rallying the people of Islam to form a common front in the face of Moscow and imposing their demands on it. Uzbeks and Kazakhs thus find themselves competing. Here again, however, the attempts at pan-Islamic mobilization collide with local ambitions and rivalries. Tadzhiks and Uzbeks are fighting over the area of Samarkand. Old conflicts have reemerged concerning the borders of Kazakhstan. All are agreed, however, in their common hatred of what remains of the USSR, and they increasingly view Russia as the heir of the Soviet state. The republics refuse to supply raw materials to the Community of Independent States, nor do they accept Russia's falling back on its own resources. The sporadic local conflicts are often bloody, but they sometimes give way to common positions threatening Russia and, above all, to the search for that shared identity—Islamic or Turkish—that would be (and the people of Central Asia know this) the surest way to influence Russia's choices. And, in Moscow, with its accumulating problems, leaders are alarmed about the erecting of the Islamic "barrier" in which population growth and the general unrest owing to underdevelopment are gradually sparking a general mobilization. The breakup of the USSR leaves the field free for a multitude of local firestorms and a north-south conflict to which Islam gives a particularly pronounced cultural dimension.

In less than two years, communism collapsed everywhere, without the conquerors being tempted to spill blood. They are to be lauded for their maturity and level-headedness. For all that, within the borders of what used to be the USSR, blood is being shed and flashpoints are multiplying. The causes of this trend are the national communities, not social groups or individuals. The heirs of Marx, who so disputed the importance of the national problem, suddenly discovered, after rejecting Marxism, that this is the most serious obstacle they face. They now know that the future of democracy depends on their responses to the aspirations or rebellions of the national groups. No issue is more urgent and decisive in what was the USSR and the mythical "historical community of a new type, the Soviet people, fruit of the friendship of peoples" than the one posed by these same people. The nation, which Lenin thought he had exorcised, has returned; it is wreaking vengeance for being ignored. It has proved this by destroying communism, for its collapse came about through the rebellion of the nations.

Notes

Introduction

1. A. Amalrik, *L'Union soviétique survivra-t-elle en 1984?* (Paris, 1976) published in the USSR as "Variant for a Newspaper," *Ogoniok, 9,* 1990, pp. 18–23.

Chapter 1

1. *Pravda,* February 26, 1986; and Mikhail Gorbachev, *Perestroika* (New York, 1987), p. 18.
2. *Pravda,* February 26, 1986.
3. *Pravda,* February 28, 1986.
4. E. V. Tadevosian, "Internationalizm sovietskogo mnogonatsional'nogo gosudarstva," *Voprosy filosofii, 11,* 1982, p. 28.
5. *Pravda,* November 27, 1985; January 2, 1986.
6. *Pravda,* November 27, 1985.
7. *Pravda,* November 28, 1987; and Gorbachev's report for the seventieth anniversary of the Revolution, in *Octobre et la restructuration, la révolution se poursuit,* A.P.N., 1987, p. 47, and interviews with Mikhail Gorbachev in the *Washington Post* and *Newsweek,* A.P.N., 1988.
8. See the editorial in *Pravda,* June 16, 1986; and E. Bagramov in *Pravda,* August 19, 1986.

9. On the problem of a change to a more Russian-dominated political personnel, see M. Tatu's excellent *Gorbachev* (Paris, 1987), particularly pp. 168–69.
10. A. I. Prigogin, in *Vek XX i Mir, 12,* 1988, p. 10.

Chapter 2

1. Alexandre Solzhenitsyn, *Archipelag Gulag* (Paris: Russian edition, YMCA Press, 1974), vol. 2, chap. 16.
2. V. D. Medvedev, *Andropov au pouvoir* (Paris, 1983), pp. 158–68; and V. Coulloudon, *La Maffia en U.R.S.S.* (Paris, 1990).
3. *Pravda,* December 11, 1982: "The Politburo had attracted the attention of the prosecutor of the USSR on the need to take measures to *improve* respect for socialist legality."
4. This was discussed by Gorbachev at the plenary session of 1987. See *Pravda,* November 28, 1987; and also Sokoloff, "Bandokratiia," in *Literaturnaia Gazeta, 33,* 1988, p. 12.
5. Russian speakers went from 1 million in 1970 to 50 million ten years later.
6. *Pravda,* November 1, 1983, and December 23, 1983.
7. *Pravda,* February 28, 1986 (interview with Usman Khodzhaev), and *Pravda Vostoka,* June 7, 1986 (annulment of the decree of December 1983 honoring Rashidov).
8. *Pravda,* April 2, 1987, and April 25, 1987, where it is emphasized that the Uzbeks are unable to cut down on corruption.
9. D. Likhanov, in *Ogoniok, 1–4,* 1989, pp. 25–30, 28–31, and 18–23, and *Ogoniok, 29,* 1988, pp. 20–23, on Rashidov.
10. *Moskovskie Novosti, 14,* April 3, 1988 (investigation by Gdlian).
11. *Literaturnaia Gazeta,* June 10, 1987.
12. A. Nekritch, *Nakasannye Narody* (New York, 1978).
13. *Literaturnaia Gazeta,* August 12, 1987.
14. *Pravda Vostoka,* July 14, 1985.
15. In the *Pravda* of June 7, 1985, Usubaliev pillories the misappropriations of funds in Kirghizia.
16. *Pravda,* July 22, 1985, where the situation in Kazakhstan is criticized.
17. *Pravda,* February 9, 1986, against Kunaev.
18. *Izvestia,* September 7, 1986.
19. *Narodnoe Khoziaistvo SSSR, 1922–1981* (Moscow, 1982), p. 33.
20. *Natsional'nyi sostav naseleniia* (Moscow) "Finansy i statistika," II, p. 68, quoted later.
21. In June, the region of Taldy-Kurgan, with a mostly Kazakh population, under Russian authority, *Natsional'nyi sostav* . . . and *Pravda,* September 21, 1989, p. 2.

22. See Kolbin in the *Pravda* of March 10, 1987, on the measures taken to "purify" the republic.

Chapter 3

1. See Alexandre Solzhenitsyn, *L'Archipel du Goulag* (Paris: Seuil, 1975), vol. 3, pp. 434–440.
2. TASS, February 19, 1986; *Literaturnaia Gazeta,* January 14, 1987; *Argumenty i Fakty, 16,* 1987.
3. This was disputed by the Kazakhs, *Izvestia,* June 8, 1989 (Mukhtar Sakarov at the Congress of People's Deputies); on the naming of an investigating commission, *Izvestia,* November 15, 1989, and *Literaturnaia Gazeta, 51,* 1989.
4. *Komsomol'skaia Pravda,* January 7, 1987 ("hooligans"), and January 16, 1987 ("provocations").
5. *Pravda,* July 22, 1986 (praise for Kolbin) and February 9, 1986 (against Kunaev).
6. *Pravda,* January 11, 1987.
7. Table based on *Itogi vsesoiuznoi perepisi naseleniia, 1959g.,* Moscow, 1962–1983; *Itogi . . . 1970 g., Narodnoe Khoziaistvo SSSR, 1922–1982,* op. cit., p. 35, *Natsional'nyi sostav,* p. 68.
8. G. Wild, *Economies nationales,* pp. 5–20 (the source of the economic data given here was G. Wild).
9. Ibid., p. 18.
10. *Pravda,* January 21, 1987, an editorial against nationalism and public disorder.
11. *Pravda,* June 12, 1987, and July 16, 1987.
12. On the fate of this community, see *Argumenty i Fakty, 22,* 1986, p. 3.
13. *Kazakhstanskaia Pravda,* June 12, 1987.
14. *Kazakhstanskaia Pravda,* February 9, 1987.
15. *Natsional'nyi sostav,* p. 68.
16. The mutual accusations took on a particular importance at the beginning of 1987, when the Central Committee of the Kazakh Communist party accused the writer Suleimanov, who was noted for his nationalistic stands and his prestige with young people, of showing leniency in an affair of corruption and participating in the adulatory filming about Kunaev. This polemic reveals a desire to mix Kazakh nationalism and corruption. See *Kazakhstanskaia Pravda,* March 15, 1987.
17. *Naselenie SSRR,* 1988, Moscow, 1989, pp. 26–28.
18. *Kazakhstanskaia Pravda,* May 6, 1987, and *Pravda,* February 11, 1987, article by T. Esilbaev.
19. T. Aitmatov, in *Literaturnaia Gazeta,* August 13, 1986.

Chapter 4

1. A. Ter Minassian, *La république d'Armenie* (Brussels, 1989), pp. 129 and 259–62. *Dossier Karabakh. Faits et documents sur la question du Haut-Karabakh, 1918–1989* (Paris-Cambridge, 1988).
2. *Natsional'nyi sostav,* p. 80.
3. Alexandre Solzhinitsyn, *Lettre aux dirigeants de l'Union soviétique* (Paris, 1974), pp. 11, 23, 32.
4. "Opasnye goroda," *Trud,* June 3, 1989.
5. Interview with Murray Fishbach, Washington, April 25, 1990. (In 1986, a samizdat document indicated that the number of mentally retarded children had quintupled and that of cancers had quadrupled). And *Naselenie SSRR,* pp. 678 and 684.
6. *Natsional'nyi sostav,* pp. 78–79.
7. *Glasnost', 10* and *11,* 1987; *Literaturnaia Gazeta,* June 24, 1987.
8. Text published in Yerevan and Baku: *Kommunist,* February 27, 1988; *Bakinskii rabochi,* February 27, 1988.
9. *Komsomol'skaia Pravda,* February 21, 1988.
10. C. Muradian, *De Staline à Gorbatchev. Histoire d'une république soviétique, l'Arménie* (Paris, 1990), p. 431.
11. *Izvestia,* March 25, 1988.
12. *Pravda,* March 21 and 24, 1988 (resolution of the Supreme Soviet).
13. *Pravda,* June 10, 1988.
14. *Pravda,* July 20, 1988 (meeting of the Supreme Soviet), *Izvestia,* June 17 and 19, 1988 (votes of the Supreme Soviets of Armenia and Azerbaijan).
15. Published on November 25, 1988, by an Azeri literary journal as a mimeographed document in Russian (which I was unable to use) and in Azeri.
16. *Vedomosti Verkhovnogo Soveta SSSR,* 49, 1988. A helpful translation appears in *Les Institutions de L'U.R.S.S.* (Paris: P. Gélard), D.F., 1989, pp. 39–43.
17. *Trud,* January 21, 1989, declaration by Major-General Kolomintsev.
18. *Izvestia,* June 2, 1989.
19. *Moskovskie Novosti, 39,* 1989, p. 2.
20. *Izvestia,* October 9, 1989; *Komsomol'skaia Pravda,* October 12, 1989.
21. The congress took place in Yerevan on November 4 and 5, 1989, attended by 400 delegates. See Muradian, op. cit., pp. 452–55.
22. *Izvestia,* October 20, 1989: "The railroads ought not to be a means of blackmail in interethnic conflicts."
23. On January 4, despite negative reactions on the spot, the Supreme Soviet of the USSR named Vladimir Foteev president of the Control Commission.
24. *Pravda,* January 11, 1990, unconstitutionality of the Armenian and Azeri decisions.
25. *Komsomol'skaia Pravda,* January 18, 1990.

26. *Pravda,* January 16, 1990 (Karabakh) and January 20, 1990 (Baku).
27. *Izvestia,* February 8, 1990.
28. Bulletin from TASS, January 19, 1990.
29. See the testimony of Gary Kasparov: *Moskovskie Novosti, 14,* January 28, 1990, p. 9; and in the same issue, p. 8: "Baku: Cho dal'she."
30. *Radio Liberty Report on the USSR, 35,* September 1, 1989, pp. 30–31.
31. Ibid., *35,* pp. 29–32.
32. *Moskovskie Novosti,* p. 4, 1990, p. 9: "Armenia: tretiia blokada."
33. "Bakinski Sindrom," in *Moskovskie Novosti, 9,* March 4, 1990, p. 13.

Chapter 5

1. *Natsionol'nyi sostav,* pp. 71 and 73–75.
2. Ibid., p. 73.
3. *Revoliutsiia i natsional'nosti, 61,* March 1935, p. 51.
4. A. Nektrish, *Nakazannye Narody.*
5. *Sur l'origine de la confrontation* (On the Origin of the Clash), see also in the informal newspaper *Referendum,* Moscow, February 1–15, 1989, pp. 9–12, the investigation by E. Edelkhanov.
6. *Zaria Vostoka,* February 19, 1989.
7. *Zaria Vostoka,* February 25 and 26, 1989.
8. *Natsional'nyi sostav,* p. 71.
9. *Izvestia,* April 10, 1989.
10. *Referendum,* 31, April 18, 1989, the article by Galina Kornilova, and Reuters, April 10, 1989.
11. *Izvestia,* April 12, 1989.
12. The *Izvestia* and *Pravda* of April 11, 1989, emphasize the return of calm. Conversely, the *Krasnaia Zvezda* of April 11, 1989, insists on the persistence of tensions.
13. "Zakliuchenie Komissii Verkhovnogo Soveta Gruzinskoi SSSR po rassledovaniiu obstoiatel'stv imevshikh mesto 9 Aprelia 1989 g.v. gorode Tblissi," published in *Zaria Vostoka,* October 5, 1989, and the report of the Sobchak committee, "O. Sobytiiakh V. gorode Tbilissi," January 29, 1989. See also the Sobchak-Ligachev dialogue in *Ogoniok, 10,* 1990, pp. 26–27.
14. *Zaria Vostoka,* May 15, 1989.
15. *Zaria Vostoka,* April 15, 1989.
16. *Ogoniok, 11,* March 1990, p. 6.
17. Ibid., p. 5.
18. See the interview with Boris Yeltsin in *Sovetskaia Estonia,* February 10, 1990.
19. *Zaria Vostoka,* report of the Shogulidze committee, October 5, 1989. pp. 2–4.
20. *Kommunisti,* May 18, 1989.

21. See I. Rost's reporting in *Referendum,* 30; Kornilova, op. cit., p. 31; and the undated supplement to *Russkaia Mysl',* "Gruziia na stranitsakh nezavisimoi petchati."
22. *Ogoniok,* text quoted, 11, 1990, p. 6.
23. On May 30, at the Congress of People's Deputies, the Georgians demanded, in addition to the resignations of Rodionov and Nikolsky, second secretary of the Georgian Communist party, the pure and simple elimination of the post of the second Russian secretary, the "satrap from Moscow."
24. The reports of Sobchak and Shogulidze, op. cit., fully confirm it. See also *Ogoniok, 10,* 1990, pp. 26–27.
25. See the article by General Shatalin, *Krasnaia Zvezda,* April 3, 1990.
26. *Moskovskie Novosti,* 37.
27. On the arms problem in Georgia, see *Moskovskie Novosti,* July 8, 1990, p. 2. And "Patrony navynos," *Izvestia,* August 19, 1987; *Pravda,* July 26, 1990.
28. *Izvestia,* July 17 and 20, 1989.
29. *Zaria Vostoka,* September 30, 1989.
30. *Natsionnal'nyi sostav,* op. cit., pp. 3–4, 71 and 76.
31. *Zaria Vostoka,* December 2, 1989.

Chapter 6

1. *Kosmsomol'skaia Pravda,* May 12, 1989.
2. *Pravda,* June 8, 1989, and June 9, 1989.
3. *Pravda,* June 12 and 13, 1989; *Krasnaia Zvezda,* June 11, 1989.
4. *Pravda,* June 21, 1989.
5. *Izvestia,* June 23, 1989.
6. *Pravda,* June 25, 1989.
7. *Komsomol'skaia Pravda,* July 10, 1990.
8. *Izvestia,* February 15, 1990. For a general picture of events in Tadzhikistan since 1985, see *Sotsialisticheskaia Industriia,* January 18, 1989.
9. *Komsomolskaia Pravda,* June 12, 1989: "The violence is, among other things, due to drugs."
10. Declaration of General Shatalin, head of the troops of the MVD, *Sel'skaia Zhizn',* June 11, 1989.
11. *Izvestia,* January 20, 1990.
12. *Stroitel'naia Gazeta,* June 11, 1989.
13. *Natsional'nyi sostav.*
14. *Natsional'nyi sostav,* p. 70.
15. The number is inexact because they are not listed as such in the census.
16. "Uzbekistan is a paradise, compared to the ethnic situation that you will find at home!" (Gumbaridze to a delegation of Meskhes.)
17. V. Sheshko, in *Sovetskaia Etnografiia,* June 1988, pp. 5–6.

18. *Natsional'nyi sostav,* p. 92.
19. See the speech of the president of the Union of Tadzhik Writers at the plenum of the Writers' Union of the USSR, *Literaturnaia Gazeta,* March 9, 1988.
20. The Tatars of Crimea are another example of these territorial imbroglios. Their republic, which was eliminated after this deportation, was annexed to the Ukraine. In 1987, a presidential commission headed by Andrei Gromyko was created to study the problem of their return, which presupposed the restoration of an autonomous republic and hence the amputation of the Ukrainian territory. On June 9, a terse bulletin announced that the commission had rejected their demand. See *Pravda,* June 9, 1988.
21. Konstantinov, in *Pravitel'stvennyi vestnik, 29*(55), 1990, p. 12.
22. The trial of the group of anti-Soviet "right-wingers and Trotskyists," shorthand account of the debates (March 2–3, 1938). Moscow 1938. pp. 239–40, cross-examination of Faizullah Khodzhaev.
23. "Aral ugrozhaet planete," *Sotsialisticheskaia Industria,* June 20, 1989; on the connection to infant mortality and malformations of newborns, see *Literaturnaia Gazeta,* January 17, 1988, p. 2; *Ogoniok, 13,* 1988, p. 26; and *Moskovskie Novosti, 26,* 1990, p. 26.
24. *Sovetskaia Kirghizia,* June 7, 1989, quoted in R.L.-R.F.E: *Report on the USSR,* June 10, 1989, p. 18.
25. Ashirov, *Musulmanskaia propoved* (Moscow, 1978), p. 75.
26. "Kremlin orders special flights for Mecca trips," *New York Times,* April 24, 1990. And *Ogoniok, 6,* 1990 (Muph Muhammad Yussuf).
27. Interview with Kharchev, in *Nauka i religiia, 11,* 1987, pp. 21–23.
28. The consumption of drugs nearly doubled between 1985 and 1990. *Sotsialnoe razvitie i uroven' zhizni naseleniia* (Moscow, 1989), p. 145.
29. *Naseleniie SSSR,* pp. 665–66.
30. A. Tsipko, *Rodina, 2* and *3,* 1990.

Chapter 7

1. A. Migranian, in *Voprosy Filosofii,* August 1987, pp. 78–80, and Gorbachev at the plenum of January 5, 1990, *Pravda,* January 8, 1990.
2. B. Kurashvili, "Aspekty perestroiki," in *Sovetskoe Gosudarstvo i Pravo,* December 1987, p. 4.
3. Statements by national writers at the plenum of the Writers' Union of the USSR, April 27–28, 1987, *Literaturnaia Gazeta,* April 29 and May 6, 1987, *Russkaia Mysl',* March 11, 1988, p. 4.
4. *Literaturnaia Gazeta, 38,* 1988, p. 1.
5. "Vremia," television program, June 13, 1988; and "Documents: The May Baltic Assembly," Nationality Papers, Fall 1989, pp. 242–59.

6. *Pravda,* November 26, 27, and 28, 1988; *Izvestia,* December 26, 1988.
7. Polls and commentaries by M. and A. Kirch of the Academy of Sciences of Estonia—published in *Nationality Papers,* op. cit., pp. 171–77.
8. *Rodnik, 12*(24), December 1988. Unpaginated document: "Otkrytoe pis'mo russko—iazychnomu naseleniu Litvy."
9. B. Nahaylo, *R. L. Research,* October 10, 1988, and B. Nahaylo, in *R. L. Report on the USSR,* February 24, 1989, pp. 15–17.
10. In Riga, the Democratic Union of Soviet Estonia held its congress on January 28, 1989; it voted for Baltic self-determination and noninterference from Moscow in the Caucasus.
11. *Report on the USSR, 11,* 1989, pp. 24–21; *14,* 1989, pp. 25–11; and *19,* 1989, pp. 17–20.
12. *Krasnaia Zvezda,* January 6, 1989.
13. *Ogoniok, 47,* 1988, p. 31.
14. *Pravda,* July 1, 1989.
15. *Natsional'nyi sostav,* p. 62.
16. Robert Conquest, *The Harvest of Sorrow: Soviet Collectivization and the Terror-Famine* (New York: Oxford University Press, 1986); V. Barka, *Le prince jaune* (Paris, 1981), p. 364; *Sotsiologicheskie issledovaniia,* June 1990, on the census of 1937 and the effects of the famine.
17. See Gorbachev's praise of him at the time of his retirement. *Pravda,* September 30, 1989.
18. *Pravda,* September 15, 1989. Foundation of the RUKH and its platform, mimeographed document, n.p., n.d., 8 pages.
19. By the name of Stepan Bandera, who directed the Ukrainian resistance to the Sovietization of the western Ukraine.
20. *Komsomol'skaia Pravda,* September 24, 1989; *Pravda,* September 30, 1989; *Krasnaia Zvezda,* October 7, 1989.
21. Interview with de I. Pakalchuk, in *Radio Liberty Report on the USSR, 47,* October 13, 1989, pp. 27–31.
22. With 218,891 inhabitants in 1989 (258,000 in 1979), the Poles are in fifth place in Ukraine. See *Natsional'nyi sostav,* p. 61.
23. *Moskovskie Novosti, 26,* July 1, 1990, pp. 6–7 (A. Mikhadze, G. Javoronkov).
24. *Ogoniok, 49,* December 1989, p. 11, a full page devoted to the poet Solih.
25. V. Falin, *Izvestia,* August 21, 1989.
26. Fourteen of the thirty-three members of the committee voted in favor. The very liberal A. Yakovlev did not sign the document. See his interview, *Pravda,* August 18, 1989.
27. Commentaries on the "seeds of division" left by the demonstrations: *Pravda,* August 23–25, 1989, and *Krasnaia Zvezda,* September 3, 1989.
28. I. Zhukov, *Pravda,* September 2, 1989.

29. During the electoral campaign, even a newspaper as conservative as *Sovetskaia Rossiia* (February 28, 1989), was distressed that whole districts were condemned to a single candidacy.
30. *Vedomosti Verkhnogo Soveta SSSR, 49,* 1988.
31. Law of January 18, 1989, text in *Sovetskaia Estoniia,* January 22, 1989.
32. Law of January 25, 1989, text in *Sovetskaia Litvia,* January 26, 1989.
33. Law of May 5, 1989, text in *Sovetskaia Litvia,* May 7, 1989.
34. Law of September 1, 1989, *Pravda,* August 28 and 29 and September 3, 1989.
35. Hence June 24, 1990, the day of the "open border," organized the Moldavians and Romanians to mark the fiftieth anniversary of Bessarabia's annexation to the USSR.
36. *Pravda,* July 21, 1989; see V. Chebrikov, *Kommunist, 8,* 1989.

Chapter 8

1. Twentieth Conference of the CPSU. Quoted by Army General N. Popov, *Partiinaia Zhizn', 2,* 1989, pp. 61–65.
2. Ibragimbeili "Natsional'nye formirovanie," *Sovetskaia Voennaia Entsiklopediia,* vol. 5 (Moscow, 1978), p. 522; *Sovetskaia armiia, armiia druzhby narodov* (Moscow, 1955), p. 56.
3. "In their activities, the organization of the Party, territory, region, . . . are guided by the platform and statutes of the CPSU. In the republic, region . . . they proceed to the whole work of implementing the Party and organize the execution of the directives of the Central Committee of the CPSU," *Party Statutes,* chapter 5, article 45.
4. The problem concerns the use of the term *sovereignty.* The noun *sovereignty* is applied in the constitution of the USSR and defines the Soviet state (article 75). The republics have only the right to the adjective *sovereign* (State, rights, articles 76 and 81). Constitution of 1977, vocabulary unchanged in 1988 and 1990.
5. See the article by A. Chubarian, director of the Institute of World History of the Academy of Sciences of the USSR in *Izvestia,* January 1, 1989; that of V. M. Kulich, *Komsomol'skaia Pravda,* August 24, 1988; and the creation of the Investigating Committee of the congress, *Izvestia,* June 3, 1989.
6. *Sovetskaia Estoniia,* November 17, 1988; *Sovetskaia Litvia,* May 19, 1989; *Sovetskaia Latvia,* July 29, 1989.
7. *Pravda,* August 17, 1989.
8. *Sovetskaia Estoniia,* October 6, 1989.
9. This did not prevent Gorbachev from expressing serious reservations about the Baltic plans for economic autonomy at the plenum of the Central Committee of the CPSU in September of 1989, *Pravda,* September 20, 1989.

10. The text appeared in *Bakinskii Rabochi,* October 5, 1989.
11. Professor Tamaz Shogulidze presented a detailed plan for this self-determination in stages.
12. *Zaria Vostoka,* March 21, 1990.
13. Resolution voted by the congress of the Communist party of Moldavia on May 18, 1990; see *Tass,* May 19, 1990.
14. Interview with Colonel Borodin, *Soiuz, 4,* January 22 and 28, 1990, p. 6.
15. General Popov, *Parrtinaia Zhizn', 2,* 1989, pp. 61–65.
16. S. Enders Wimbush and Alex Alexeiev, *The Ethnic Factor in the Soviet Armed Forces* (Santa Monica: Rand Corporation, 1983), p. 13.
17. *Krasnaia Zvezda,* December 22, 1989, p. 2; *Komsomol'skaia Pravda,* April 13, 1989.
18. *Krasnaia Zvezda,* April 2, 1988, p. 2.
19. Yasov (general) in *Krasnaia Zvezda,* September 22, 1989, p. 2. Parfenov (major-general) in *Krasnaia Zvezda,* October 13, 1989, p. 2.
20. J. Poliakov in *Iunost', 11,* 1987, pp. 46–48. Zhakubov and Pulatov in *Ogoniok,* March 10, 1990, p. 20.
21. September 15, 1985; *Komsomol'skaia Pravda,* June 15, 1987; A. Dolgikh, "Shkola internatsionalizma i druzhby," *Agitator, 3,* February 1988, pp. 17–20; *Krasnaia Zvezda,* September 13, 1989, p. 1 (V. Moroz).
22. *Homeland,* April 4, 1990: 105 young Estonians demanded political asylum to avoid military service.
23. V. Landsbergis: "In Lithuania, we regard the Soviet forces as an army of occupation," *International Herald Tribune,* August 22, 1989.
24. *Zaria Vostoka,* June 8, 1986.
25. D. Yazov, in *Krasnaia Zvezda,* March 7, 1989, p. 2; Lizychev (general), in *Kommunist, 3,* 1989, pp. 16–17; Moiseev (general), in *Krasnaia Zvezda,* February 10, 1989, p. 1; and *Pravda,* April 16, 1989: "Voprosy kvoennym."
26. Kuzmin (general), in *Krasnaia Zvezda,* March 16, 1989, p. 6; and *Krasnaia Zvezda,* July 26, 1989, p. 2.
27. Interview with Colonel Durnev, *Argumenty i Fakty, 8,* 1990.
28. *Moskovskie Novosti,* March 18, 1990, p. 1, federation-wide poll of 2,896 persons.
29. *Pravda,* September 20, 1989.
30. It was this battle against the federalist leanings of his colleagues from the Caucasus that linked him with Stalin in 1913. He assigned him to the *Marxism and the National Question* published in two installments in *Prosveshchenie.*
31. The next day, the delegates to the congress who had voted against independence founded the Lithuanian Organization of the CPSU, a dissident authority of which little was subsequently heard.
32. *Novosti,* January 16, 1990.
33. *Pravda,* January 12–15, 1990.

34. *Izvestia,* January 16, 1990.
35. See the statement made by I. Reshetov, in charge of humanitarian problems at the Ministry for Foreign Affairs, *Izvestia,* March 28, 1990.
36. *Tass,* April 13, 1990.
37. Interview with the BBC, quoted by *R.P.E.-Radio Liberty Daily Report, 79,* April 24, 1990.
38. See statement by Colonel Petrushenko to the Congress of People's Deputies, *Izvestia,* March 17, 1990.
39. In return, the Estonians threatened to demand the region of Petseri, located in the southeast of the republic, which in 1945 was annexed by Russia because of its large Russian population.
40. The *Izvestia* of April 30, 1990, had published the results of a poll on independence taken in Latvia: 92 percent of the Latvians and 45 percent of the residents in Latvia said they favored secession.
41. The Council of Baltic States had been created by a treaty signed in Riga on September 12, 1934, and had given the Balts representation at the League of Nations, where Latvia represented the three states.
42. See K. Mihailisko, "For Our Freedom and Yours: Support among Slavs for Baltic Independence," *Radio Liberty Report on the USSR,* May 25, 1990, pp. 17–18.
43. See in the same issue, p. 16, Y. Aslan, "Muslim Support for Baltic Independence."
44. *Natsional'nyi sostav,* p. 19.
45. That is how the Ukrainian nationalists, like the writer Ivan Drach, one of the founders of RUKH, expressed themselves. See *Literaturnaia Gazeta, 15,* 1990, p. 11.
46. Interview with V. Landsbergis, *Moskovskie Novosti, 26,* 1990, p. 7.
47. *Pravda,* July 10, 1990, p. 1, decree forming the delegation of the USSR to the negotiations: eleven persons including Prime Minister Ryzhkov.

Chapter 9

1. On Russian nationalism, see John B. Dunlop, *The Faces of Contemporary Russian Nationalism* (Princeton, N.J.: Princeton University Press, 1983), and *The New Russian Nationalism* (New York: Praeger, 1985).
2. *Sovetskaia Rossiia,* January 3, 1986.
3. Maxim Gorki, "On the Russian Peasantry," in *SSSR vnutrennie protivorechiia* (Tchalidzé Publications, 1987), p. 218.
4. I. Belov, *Kanuny* (French edition: *Veilles*) (Paris, 1985), p. 420.
5. Letter published in *Ogoniok, 51,* 1986, p. 11, cosigned by Bondarev.
6. Interviews with Murray Feshbach, Washington, D.C., April 23–25, 1990. Information taken from a forthcoming book by him.

7. G. Litvinova, *Svet i teni progressa* (Moscow, 1989), pp. 251–52. And I. Bromlei, *Pravda,* February 13, 1987.
8. *Argumenty i Fakty, 33,* 1987. Quoted by G. Litvinova, op. cit., p. 252.
9. G. Litvinova, op. cit., p. 269 and I. Bromlei, *Pravda,* February 13, 1987; and V. Tishkov, in *Kommunist, 1,* 1989, p. 54.
10. Mikhail Gorbachev, in *Pravda,* January 28, 1987.
11. G. Litvinova, op. cit., p. 253.
12. *Ogoniok,* 41, 1988, p. 1. Letter to the editor.
13. S. Averintsev, "Vizantiia i rus'. Dva tipa dukhovnosti," *Novy mir, 3,* 1988, pp. 210–11; and *Literaturnaia Gazeta,* August 3, 1988, p. 7, on this plan.
14. *Pravda,* April 30, 1988.
15. A poll taken by the Institute of Sociology of the Academy of Sciences of the USSR inquired whether the prerevolutionary names of the cities and streets should be restored. Affirmative answers went from 46 percent to 77 percent. See op. cit., p. 44, table 41, and an interview with Anatoly Sobchak, "Bez diktatury," in *Ogoniok, 18,* 1990, p. 3.
16. See the statements of Averintsev in *Druzhba narodov,* June 1988, pp. 245–62, and of Alla Latynina, *Novy mir,* August 1988, p. 240.
17. Quoted by D. Hammer, *Glasnost and the Russian Idea,* Radio Liberty, in *Russian Nationalism Today,* December 19, 1988, p. 15.
18. See *Moskovskie Novosti,* March 4, 1990, p. 16, and the interviews with the American historian Stephen Cohen concerning his biography of Bukharin.
19. *Nash Sovremennik, 4,* 1988, pp. 160–75, and his argument with B. Sarrov in *Literaturnaia Gazeta, 10,* 1989, p. 2.
20. Alla Latynina, "Kolokol'nyi zvon . . . ne molitva," *Novy mir,* August 1988, pp. 232–45.
21. Ibid., p. 24, and the debate with S. Chuprinin, *Literaturnaia Gazeta, 14* and *17,* 1989.
22. *Moskovskie Novosti,* August 1, 1988.
23. N. Andreieva, *Sovetskaia Rossiia,* March 13, 1988, p. 3. See also the article by Prokhanov, *Literaturnaia Gazeta,* August 6, 1988.
24. On the movement's development, see the article by Anishchenko, "Kto vinovat," *Glasnost,* 15.
25. Tsipko, "Russkie ukhodiat iz rossii?," *Izvestia,* May 16, 1990, p. 1; "Istoki stalinizma," in *Nauka i zhizn', 11* and *12,* 1988, and *1* and *2,* 1989. And "Neobhodimo potriasenie mysli," *Moskovskie Novosti, 16,* 1990, p. 3.
26. As well as Yeltsin's "five hundred days," see the interview of the vice-president of the Soviet of the RSFSR, R. Khasbulatov, *Moskovskie Novosti, 28,* 1990, p. 10.
27. When questioned by the *Literaturnaia Gazeta* after his nomination to the council, about his contribution (the letter from seventy-four writers) to a highly anti-Semitic text, Rasputin declared that the statements in it were excessive and unfortunate.

28. *Sovetskaia Rossiia,* 8 and 9, September 1989 (on the front).
29. *Literaturnaia Gazeta,* June 13, 1990, p. 2; *Moskovskie Novosti, 24,* 1990, p. 4; *Argumenty i Fakty,* 24, 1990, p. 5.
30. *Kommunist, 15,* 1986, p. 17.
31. "Esli my pridem k vlasti," N. Andreieva, *Argumenty i Fakty, 12,* 1990, p. 4.
32. John B. Dunlop, "The Contemporary Russian Nationalist Spectrum," in *Russian Nationalism Today,* p. 9.
33. Electoral law, *Sovetskaia Rossiia,* November 1 and 3, 1989.
34. Interview with N. Travkin, *Argumenty i Fakty, 8,* 1990, p. 4.
35. *Literaturnia Gazeta, 52,* 1990, pp. 2–3.
36. *Pravda,* March 25, 1990; *Sovetskaia Rossiia,* March 28, 1990, *Trud,* March 14, 1990, pp. 4–6; and a breakdown of the votes of the first session, *Argumenty i Fakty, 26,* pp. 4–6.
37. Vlasov withdrew on May 25, and then returned to the competition because Polozkov, who succeeded him, did not manage to beat Yeltsin in the first round.
38. "Dva Pretendenta, dve programmy," *Izvestia,* May 26, 1990.
39. *Moskovskie Novosti, 26,* 1990, p. 2, and G. Popov, *Ogoniok, 10,* 1990, pp. 4–5.
40. Declaration on the sovereignty of Russia, June 12, 1990, *Argumenty i Fakty, 24,* 1990, p. 1. Decree on the government in Russia, *Argumenty i Fakty, 25,* 1990, p. 1. And on multipartyism, article 6 of the Russian constitution, adopted by 788 deputies against 32, and 17 abstentions. *Argumenty i Fakty, 25,* 1990, p. 2.
41. *Ogoniok, 8,* 1990, p. 5, and Yegorov, *Moskovskie Novosti,* September 1990, p. 14.
42. On the political parties: *Moskovskie Novosti, 28,* 1990, pp. 8 and 9.
43. A. Tsipko, *Izvestia,* May 26, 1990, p. 1.
44. A. Tsipko, *Moskovskie Novosti, 24,* 1990, p. 6.
45. G. Nivat, in *La Lettre internationale,* Spring 1990, p. 24.

Chapter 10

1. Round Table of *Voprosy Filosofii,* September 1988; discussion in *Vek XX i Mir, 12,* 1988.
2. *Vek XX i Mir,* op. cit., pp. 10–13.
3. *Izvestia,* October 22, 1948 (plan).
4. Report of the president of the Supreme Soviet of Estonia, *Sovetskaia Estoniia,* November 17, 1988.
5. Ibid.
6. *Izvestia,* December 3, 1988 (final text).
7. After the constitutional revision of March 14, 1990, it became article 124.
8. *Pravda,* November 12, 1988 (debate on the "national plenum").

9. "Natsional'naia politika Partii v sovremennykh usloviiakhy," *Pravda,* August 17, 1989, and *Pravda,* July 16, 1989.
10. "Stepen Svobody," *Ogoniok, 31,* 1989, pp. 26–27.
11. The Russian lexicon's only word for sovereignty is *Suverenitet.*
12. *Pravda,* August 19, 1989.
13. *Pravda,* September 20, 1989. Plenum: *Pravda,* September 22 and 24, 1989, p. 1.
14. *Vek XX i Mir,* March 1989, pp. 10ff.
15. Personal interview with G. Tarazevich at the Supreme Soviet, November 2, 1989.
16. The Abalkin plan was opposed by the Supreme Soviet at the end of 1989. For Abalkin, the federation could not survive without preserving a certain ownership by the union in certain key sectors.
17. *Pravda,* February 6, 1990.
18. I. Muksinov, in *Sovetskoe Gosudarstvo i Pravo,* October 1990.
19. Plan in *Pravda,* March 6, 1990, text approved by the congress in *Pravda,* March 6, 1990.
20. Press conference, Paris, March 11, 1990 (on the premises of the Calmann-Lévy publishers). Interview in *Argumenty i Fakty, 9,* 1990, p. 4, "ia vse taki optimist." On confidence: N. Popov in *Ogoniok, 7,* 1990, pp. 2–5.
21. Report of B. Gidaspov, president of the Commission of Mandates, *Izvestia,* May 26, 1989.
22. *Pravda,* April 1, 1990.
23. *Pravda,* August 4, and June 10, 1989.
24. *Izvestia,* June 10, 1989, p. 2.
25. Article 127-4 of the law of March 14, 1990.
26. Composition of the presidential council, *Izvestia,* March 25, 1990, p. 1.
27. *Literaturnaia Gazeta, 14,* 1990, p. 1.
28. Tadevosian, in *Kommunist, 6,* 1990.

Chapter 11

1. *Pravda Vostoka,* November 25, 1989.
2. *XXII s'ezd Kommunisticheskoi Partii Sovestokogo Soiuza* (Moscow, 1961), pp. 362 and 402.
3. Table compiled on the basis of the censuses of 1959, 1970, 1979, and *Natsional'nyi sostav,* pp. 3, 5, 59, 62, 64, 68, 71, 77, 81, 83, 85, 87, 90, 93, 94, 96, and 97.
4. See, for example, D. Critchlow, in *Radio Liberty Report on the USSR,* May 18, 1990, pp. 8–9.
5. Analyzed by A. Sheehy, *Radio Liberty Report on the USSR,* February 4, 1989, p. 2.
6. *Natsional'nyi sostav,* pp. 97–98, and *Naselenie SSSR,* pp. 106–08.

7. Ibid., p. 97.
8. See Girenko in *Pravda,* November 30, 1989, who emphasizes the difficulty in restoring a Tatar state.
9. "Osh, nasha tragediia," *Moskovskie Novosti, 24,* 1990, pp. 1 and 5.
10. *Izvestia,* August 25, 1989.
11. See N. Gellert's statement in *Pravda,* September 20, 1989, at the "national plenum" on the territorial problem of the Germans. And *Moskovskie Novosti, 24,* 1990, 11.
12. *Natsional'nyi sostav.* It is to be noted that next to the heading "Jews," to which these figures correspond, we find on the same page nonexistent labels in the past: "Jews of the Mountain" (probably Daghestan): 9,389 in 1970 and 19,516 in 1989; Georgian Jews: 8,455 in 1979, and 16,123 in 1989, and Jews of Central Asia: 28,308 in 1979 and 36,568 in 1989.
13. *Natsional'nyi sostav,* p. 40.
14. "Chernobyl sovershenno Sekretno," *Oppozitsia* (published in Tartu), *1*(7) January 1990, pp. 5 and 8.
15. *Moskovskie Novosti, 5,* 1990, p. 4; and *26,* 1990, p. 6; *Ogoniok, 52,* 1989, p. 5, letter from a reader in Chimkent.
16. See in *Ogoniok, 28,* 1990, p. 4, a readers' proposal to deposit the contributions of the Communist party to a refugee aid fund.
17. *Moskovskie Novosti, 26,* 1990, p. 6, results of a poll taken in Moscow in which 56 percent of the inhabitants did not want any refugees.
18. Ibid.
19. *Intelligentsia o sotsial'no-politicheskoi situatsii v strane, op. cit.,* 43, table 39. The national problem preexisted glasnost for the majority of the people polled.

Conclusion

1. *Sovetskii narod. Novaia istoricheskaia obshchnost' liudei* (Moscow 1975), 520 pp. See the bibliography on the Soviet people and the friendship of peoples.
2. *Izvestia,* July 24, 1990, p. 2; *Ogoniok, 28,* 1990, p. 4, letter from the soldier Avilov.
3. Debate between Travkin and Amsbartsumov, *Moskovskie Novosti, 11,* 1990; Burlatsky-Shmelev interview, *Literaturnaia Gazeta, 29,* 1990, p. 3.
4. *Pravitel'stvennyi Vestnik,* 28, 1990, p. 3.
5. The Supreme Soviet of the USSR's Committee for the Defense and Security of the State held hearings on the murder of the officers: *Krasnaia Zvezda,* June 13, 1990, p. 1.
6. To the question, "To settle the national conflicts in a republic, should a state of siege be declared?" the majority of those polled from all social classes answered in the negative (an average of 60 percent). *Intelligentsia o sotsial'no-politicheskoi situatsii v strane,* p. 43, table 40.

Bibliography

This book, written about a recent and troubled period, is based essentially on Soviet publications. The bibliographical items mentioned in the notes are not repeated here; only a few recent sources and publications are noted.

Main Newspapers and Journals Consulted

Pravda
Izvestia
Krasnaia Zvezda
Komsomolskaia Pravda
Sovetskaia Estonia
Zaria vostoka
Kazakhstanskaia Pravda
Bakinskii rabochii
Sovetskaia Rossiia
Literaturnaia Gazeta
Moskovskie Novosti
Argumenty i Fakty
Pravitel'stvennyi vestnik
Ogoniok
Novy mir

Kommunist
Izvestia TsK KPSS
Voprosy istorii KPSS
Oktiabr
Nash sovremennik

Reference Books: Directories and Yearbooks

Natsional'nyi sostav naseleniia (Moscow, 1989), pt. II.
Naselenie SSSR, 1987. Statisticheskii spravochnik (Moscow, 1988).
Naselenie SSSR, 1988. Statisticheskii ezhegodnik (Moscow, 1989).
Slovar' natsional'nosteii i iazykov (Moscow, 1988).
Narodnoe Khoziaistvo SSSR, 1922–1982 (Moscow, 1982).
Narodnoe Khoziaistvo SSSR, 1983–1989 (Moscow).
Narodnoe Khoziaistvo SSSR, Za 70 let (Moscow, 1987).
Narodnoe obrazovanie i kultura v SSSR (Moscow, 1989.
SSSR, Administrativno-territorial'noe delenie soiuznykh respublik (Moscow, 1987).
Materialy XXVII s'ezda KPSS (Moscow, 1986).
Materialy plenuma TsK KPSS, 27–28 ianv. 1987 (Moscow, 1987).
Materialy XIX vsesoiuznoi konferentsii. Kommunischeskoi partii sovetskogo soiuza i stenograficheskii ochet (Moscow, 1988), 2 vols.
Narodnyi Kongress. Sbornik materialov Kongressa narodnogo fronta Estonii 1.–2. Okt. 1988 (Tallin, 1989).
Gorbachev, Mikhail. *Perestroika: New Thinking for Our Country and the World* (New York: Harper & Row, 1987).
Yeltsin, Boris. *Against the Grain: An Autobiography,* trans. Michael Glenny (New York: Summit, 1990).

Books about Gorbachev's USSR

Kerblay, Basile H. *Gorbachev's Russia* (New York: Pantheon, 1989).
Marie, Nadine. *Le Droit retrouvé: Essai sur les droits de l'homme en U.R.S.S.*) (Paris: Presses Universitaires de France, 1989).
Medvedev, Zhores A. *Gorbachev* (New York: Norton, 1986).
Muraka, Dev. *Gorbachev: The Limits of Power* (London: Hutchinson, 1988).
Nove, Alec. *Glasnost' in Action: Cultural Renaissance in Russia* (Boston: Unwin Hyman, 1989).
Romano Sergio. *La Russia in Bilico* (Bologna: Il Mulino, 1989).
Tatu, Michel. *Gorbachev: The Origins of Perestroika* (Boulder, Colo.: East European Monographs, 1991).
Thom, Françoise. *Le Moment Gorbatchev* (Paris: Hachette, 1989).

Recent Studies of the National Problems

Hérodote. "Géopolitique de l'U.R.S.S.," 47; "Les marches de l'U.R.S.S.," 54.

Radvanyi, I. *L'U.R.S.S: régions et nations* (Paris, 1990).

Alexeiev, Alexander R., and S. Enders Wimbush, eds. *Ethnic Minorities in the Red Army: Asset or Liability?* (Boulder, Colo.: Westview, 1988).

Bialer, Seweryn, ed. *Politics, Society, and Nationality inside Gorbachev's Russia* (Boulder, Colo.: Westview, 1989).

Bilinsky, J. *A Successful Perestroika in Nationality Relations* (New York: forthcoming).

Conquest, Robert, ed. *The Last Empire: Nationality and the Soviet Future* (Stanford, Calif.: Hoover Institution, 1986).

Enloe, Cynthia K. *Police, Military, and Ethnicity: Foundations of State Power* (New Brunswick, N.J.: Transaction Books, 1980).

Karklins, Rasma. *Ethnic Relations in the U.S.S.R.: The Perspective from Below* (Boston: Allen & Unwin, 1986).

Motyl, Alexander J. *Will the Non-Russians Rebel? State, Ethnicity, and Stability in the USSR* (Ithaca: Cornell University Press, 1987).

Mouradian, Claire. *De Staline à Gorbatchev: Histoire d'une république soviétique: L'Arménie* (Paris: Ramsay, 1990).

Rywkin, Michael. *Moscow's Muslim Challenge: Soviet Central Asia* (Armonk, N.Y.: M. E. Sharpe, 1990).

Ter Minassian, Anahide. *La République d'Arménie* (Brussels: Editions Complexe, 1989).

Wizniewska, I. *Paroles Dégelées. Ces Lituaniens qu'on disait soviétiques* (Paris, 1990).

———. *L'Eglise en Ukraine. De la contrainte à la liberté* (Paris: Istina, 1989).

Concerning the Nation

Armstrong, John Alexander. *Nations before Nationalism* (Chapel Hill: University of North Carolina Press, 1982).

Bromlei, Iulian F. *Ethnography and Ethnic Processes* (Moscow: Social Science Today Editorial Board, USSR Academy of Sciences, 1978).

———. *Etnosotial'nye protsessy i teoriia, istoriia, sovremennost'* (Moscow: 1987).

Deutsch, Karl Wolfgang. *Nationalism and Social Communication* (Cambridge: MIT Press, 1966).

Gellner, Ernest. *Nations and Nationalism* (Oxford: Blackwell, 1983).

Seton-Watson, Hugh. *Nations and States: An Enquiry into the Origins of Nations and the Politics of Nationalism* (Boulder, Colo.: Westview, 1977).

Index

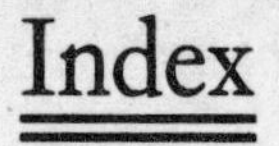